T0283967

Wisconsin
FARMS AND
FARMERS MARKETS

Tours, Trails *and* Attractions

KRISTINE HANSEN

Globe
Pequot

Guilford, Connecticut

Globe
Pequot

An imprint of The Rowman & Littlefield Publishing Group, Inc.

4501 Forbes Blvd., Ste. 200

Lanham, MD 20706

www.rowman.com

Distributed by NATIONAL BOOK NETWORK

British Library Cataloguing in Publication Information available

Library of Congress Cataloging-in-Publication Data available

ISBN 978-1-4930-5581-4 (paper : alk. paper)

ISBN 978-1-4930-5582-1 (electronic)

♾™ The paper used in this publication meets the minimum requirements of American National Standard for Information Sciences—Permanence of Paper for Printed Library Materials, ANSI/NISO Z39.48-1992.

CONTENTS

INTRODUCTION

I have yearned to write this book for years. Ever since I shared a bus seat with celebrity chef Rick Bayless on a Willy Street Co-op–sponsored farm tour. While I listened to Rick and other Chicago chefs discuss restaurant menus with farmers, Wisconsin's deep farm-to-table connection unfurled right in front of me, bridging the gap between rural culture and the bright lights of a big city's restaurant culture.

As a workshare volunteer at Pinehold Gardens and Wellspring Organic Farm and Retreat Center (both are near Milwaukee), I connected myself—quite literally—to the soil for four hours a week. My gift? A week's worth of fresh, in-season produce, much of which was new to me (like green-zebra tomatoes or knobby kohlrabi heads) and all of which taught me to be a better cook.

Wisconsin ranks second to California in the number of organic farms, earning national acclaim for commercial crops (ginseng, cranberries, and horseradish) while also producing fruits, meats, cheese, and vegetables on a boutique scale. Many farmers are young—working with dirt is their dream job. They bring other skills in art, cooking, marketing, and business. Some are reviving the family farm, honoring roots as far back as seven generations. Still others arrive from other states, enticed by land prices, quality of the soil, and an audience eager for—and by now expecting—farm-to-table food.

Whether it is picking strawberries in June, eating at a pizza farm, experiencing goat yoga at a creamery, sipping wine at a vineyard, scooping up cherries at a farm stand to bake a pie, stepping foot into a cranberry bog or fishing for trout, Wisconsin is no stranger to agritourism. There are numerous ways to experience farms in their natural element.

If you love food and supporting small businesses—this book is for you. Tuck it into your backpack or bag as you travel around the state and let your palate be your guide. I have even included overnight stays on farms in case you are not ready to go home just yet.

Working on this book came with challenges. The COVID-19 pandemic canceled research trips. County fairs were also cancelled, farm-stays stopped accepting bookings, pizza farms went on pause, and attractions shuttered. As restrictions lifted, I found solace in Wisconsin's countryside, transported far, far away from my Milwaukee home as I bought exotic mushrooms at River Valley Ranch or met Peru's most beloved animal at an alpaca ranch. My first meal "out" was a "Grazer" pizza at Grassway Organics in East Troy, on an unusually warm Saturday night in early May.

Through factchecking and interviews, I have confirmed details on visiting each farm. But it is still best to call first. You would not want to pull up to a locked gate, now would you?

A Guide to Seasonal Fruits and Vegetables in Wisconsin

If you are a frequent shopper at farmers markets you already know that what is sold at stands shifts each month, based on that year's harvest. Thankfully, there are some consistent bounties every year, starting with beets in April and ending with pumpkins in November. Here is a handy guide to help plan your shopping.

May: Arugula, asparagus, cucumbers, garlic scapes, parsnips, potatoes, radishes, salad mix, and spinach

June: Arugula, asparagus, beets, bok choy, broccoli, cabbage, cucumbers, collard greens, garlic scapes, kale, kohlrabi, lettuce, sweet peas, potatoes, radishes, raspberries, rhubarb, salad mix, scallions, spinach, strawberries, sweet cherries, and turnips

July: Apples, arugula, beans, beets, bell peppers, blueberries, bok choy, broccoli, cabbage, carrots, cauliflower, collard greens, corn, cucumbers, eggplant, kale, kohlrabi, lettuce, okra, scallions, summer squash, sweet cherries, Swiss chard, tart cherries, melons, raspberries, tomatillos, tomatoes, and turnips

August: Apples, arugula, beans, beets, bell peppers, bok choy, blueberries, broccoli, collard greens, corn, cucumbers, eggplant, garlic, grapes, hot peppers, kale, kohlrabi, leeks, melons, okra, onions, pears, potatoes, raspberries, scallions, Swiss chard, tomatillos, tomatoes, and winter squash

September: Apples, arugula, beans, bell peppers, Brussels sprouts, cabbage, carrots, cauliflower, collard greens, corn, cranberries, cucumbers, eggplant, garlic, grapes, hot peppers, kale, kohlrabi, leeks, okra, onions, pears, potatoes, radishes, raspberries, rutabagas, summer squash, Swiss chard, tomatillos, tomatoes, and winter squash

October: Apples, arugula, beets, bell peppers, bok choy, broccoli, Brussels sprouts, cabbage, carrots, cauliflower, collard greens, cranberries, cucumbers, eggplant, garlic, kale, kohlrabi, leeks, lettuce, onions, parsnips, potatoes, radishes, raspberries, rutabagas, scallions, spinach, Swiss chard, winter squash, tomatillos, and turnips

Year-round: dry beans, hickory nuts, honey, maple syrup, microgreens, mushrooms, and sorghum

Source: UW-Madison Center for Integrated Agricultural Systems, cias.wisc.edu

Today's Farmers Market Vendor

The ages-old image of a farmer—just like Iowan painter Grant Wood's *American Gothic* of an elderly male in denim overalls—has been replaced. Nowhere is that truer than at your local farmers market. It is here you will see, in full view, how different cultural and ethnic groups around the state have made a career out of farming, baking, gardening or another artisan-food craft.

I can *parlez en francais* with a French nun ordering macarons at the Kenosha HarborMarket, bite into a Belgian-style waffle baked to a crisp by a young couple inside their trailer parked at Milwaukee's South Shore Market, or get tips for an upcoming Greece vacation from Mavra Papadatos of Mavra's Greek Olive Oil at the Greenfield Farmers Market. Mavra grew up on a farm in Amalias Elias, Peloponnisos, Greece, where the olives to craft her oil are grown and then pressed. The more you shop these markets, the clearer it becomes that the "meat and potatoes" assumption—and our crushes on cheese—the coasts have about the Midwest are so not true. We like to eat eclectically here—even more if we cook it ourselves, folding in farm-fresh ingredients.

These are only the cultural experiences at markets near my home in Southeastern Wisconsin, of course. If I travel farther west, into Green County or the Driftless Region, Swiss roots will likely play into a farmer's genetic make-up and I may very well find alpine-style cheeses for sale, too. The vendors who set up shop at Madison's Capitol Square each Saturday morning may be Mennonite bakers or cheesemakers who learned the craft from their Swiss father (Willi Lehner of Bleu Mont Dairy).

Throughout the state is a rich tapestry of Hmong farmers continuing their family's tradition selling alongside Caucasian couples who put everything on the table to pursue their farming dreams. They seed, sow, and till in tandem each spring, working vigorously all summer before winding down with the last root crops in late November.

11 Ways to Support Small Farms and Farmers

Shop at local farmers markets.

There is not a city, town, or village in Wisconsin without a farmers market or access to one within a few miles. I was shocked to find this out when writing this book thinking, surely there must be a spot where locals have to drive an hour to get to a farmers market. Nope. This is the easiest, and most fun, thing you can do.

Sign up for their email newsletters.

It is important to know the face behind the farm—or, the voice behind the email newsletter. As a subscriber, you will stay informed about not only what is fresh and in season, but also fun events at the farm featuring live music, pizza, chef dinners, you-pick dates, and what is new at the farm stand or store.

Follow them on social media (especially Facebook and Instagram).

Farmers are out in the fields daily planting seeds and picking crops, with rarely enough time to update a website. Adding a quick note to their Facebook page about strawberry-picking dates or a photo to Instagram of what will be at that week's farmers market is easier on them—and still allows you to stay in touch.

Sign up for a CSA.

Often referred to as a CSA, community-supported-agriculture programs keep farmers afloat by locking in their income each spring via weekly (or twice-monthly) shares of fruits and vegetables to their customers. In recent years, farms are taking a curated approach by including items they do not grow, make, or sell but are still local.

Gift with art.

Farmers in Wisconsin need something to do in the off season other than order seeds and feed the animals, and unless they are wintering in Florida they likely have time to kill. This is when many indulge in their creative side hustles in painting, knitting, basketry, or woodworking. You will find these goods for sale through their website or, once the farm starts up again for the season, at their farm stands. Is there a better gift for your friend, co-worker, or family member than something homemade?

Eat at farm-to-table restaurants.

This is where Wisconsin truly excels. Farm-to-fork, farm-to-table, or whatever you choose to call it has been humming along with great success in the state, whether in large cities like Madison or Milwaukee, or homey cafes in the Driftless Region and Door County. Chefs want to work with local farmers because what they grow is so much fresher than anything coming in by plane, train, or truck. And because some farmers are literally dropping off the vegetables, fruits, and meats themselves at the restaurant, the time it takes from harvesting to cooking can be as little as twelve hours. What is picked that morning may be on that night's dinner menu.

Choose local fish.

In my research, I found two farms that not only raise fish but will—for a small fee—hand you a fishing pole and let you do it yourself. (You can do this at Rushing Waters Fisheries, in Palmyra.) Yes, we know that Maine lobster and Alaskan salmon are delicacies, but so are farm-raised trout. Then, in Door County, you would be a fool to not order whitefish, whether it is smoked spread or a fish boil.

Drink local wine.

It is easy to be a beverage snob and only drink Napa Valley Cabernets or Provence Rosé, but who is this helping other than your local liquor store? It is time to stop rolling our eyes at Wisconsin wines because the quality has improved in recent years, so much that you can find options like Wollersheim Winery's Prairie Fumé on restaurant menus around the state. And the local-wine section at big-box retailers like Woodman's is almost as spacious as Portugal's.

Buy a Christmas tree.

You have two options every November: pull that faux tree up from the basement or wrestle a fresh one onto your car. Even if you cannot get out to a farm to cut your own, Christmas-tree farmers come to you, bringing their Balsam, Spruce, and Fraser firs to lots in the city. You will not regret this move the minute you take your first whiff of fresh fir, a scent that no candle can duplicate quite as well.

Spend the night at a farm.

Skip the chain hotel and, yes, even the cool boutique hotel, to be a farmer for a night. Some Wisconsin farms have added overnight accommodations that let you, as little or as much as you like, experience life on the farm, whether that is near a cranberry bog, inside a tiny house surrounded by vegetable gardens, or with adorable goats or alpacas running around.

Go berry picking.

I have saved the best for last. Here in Wisconsin we make the most of that slim season called summer and that could mean picking berries: strawberries in June, raspberries and cherries in July,

and blueberries in August (with shifts each year thanks to Mother Nature's weather patterns). Around the state you will find numerous options for you-pick farms (just be sure to follow their websites and social media for news of pick dates) although cherries tend to grow best and most prolifically in Door County.

SOUTHEAST WISCONSIN

Despite its urban density, Milwaukee is where you are most likely to meet a farmer. Pick a day of the week and there is a different farmers market. Most farmers staff the stands themselves. Menus at farm-to-table restaurants like Braise, Sanford, and La Merenda provide a taste of farm life that is just a short bicycle or bus ride away. Shoppers at Outpost Natural Foods Co-op and Riverwest Co-op fill carts with locally grown corn, tomatoes, lettuce, and more.

Beyond Milwaukee, Walworth County is dotted with farm stands like The Elegant Farmer (try the "pie baked in a brown paper bag," seriously) and River Valley Ranch & Kitchen (fungi fans, this is your utopia), plus a darling creamery near Lake Geneva (Highfield Farm Creamery) with the cutest cows and gingham-check curtains you will ever see. In Milwaukee's northern suburbs, farm stands like Schmit's Farm Produce and you-pick orchards provide an easy escape from the city. Near the state line, Jerry Smith Farm's Green Acres is a popular spot each fall while all the cool kids hang out at Grassway Organics for pizza and live music weekend nights during the summer. Two cider producers (Armstrong Apples, Orchard & Winery and Aeppeltreow Winery) also welcome you to the farm and to stroll in the orchard, while vineyards just around Milwaukee provide a refreshing respite from beer. (No worries if you are a beer drinker: Duesterbeck's brewery-on-a-farm concept in Elkhorn has you covered.)

But you need not leave the city to pluck a tomato from the vine. Alice's Urban Farm and Community Garden is smack-dab in the middle of Milwaukee and hosts biodiversity walks, plant and vegetable sales, and outdoor yoga. Walk the labyrinth to feel some Zen.

And who can forget the state's largest fair? Each August foodies and animal lovers head to Wisconsin State Fair for cream puffs and glimpses of cows, sheep, rabbits, goats, and more.

Highfield Farm Creamery

W4848 State Line Rd., Walworth; 262-275-3027; highfieldfarm.com

Denise and Terry Woods operate the state's smallest marketing parlor—along with a farmstead that's home to Jersey cows and a cute store with gingham-check curtains. It is a dream second-chapter career after living in California and returning home (for Denise) to the Midwest. In fact, Denise signs all the creamery's Facebook posts from "the Quiet Side of the Lake," a reference to the bustling, resort town of Lake Geneva nearby.

Although they moved to their farm during the 1980s, it was not until 2014 that they began to make cheese. Terry crafts the cheese in small batches while Denise handles all the marketing and behind-the-scenes administrative work. She also is responsible for the store's aesthetic. On a visit you can see cheese (cave-aged cheddar, the bloomy-rind "Village Square" and "Roundabout," and cheese curds) being made by peering through a glass wall. Terry learned to make cheese in Scotland and also while interning at the University of Wisconsin-Madison's creamery, followed by stints observing the masters at Wisconsin's own Uplands Cheese Company and Cedar Grove Creamery. In 2015 this former computer engineer became a licensed cheesemaker.

The couple's love for antique cars shines through in their thoughtful renovation of a 1930 Model A Ford Panel Delivery Truck—you just may spot it at a farmers market or out on delivery. Denise is passionate about researching local history and that includes their 1911 barn and 1840 farmhouse. She was thrilled to learn a cheesemaker once lived on their property, and they are continuing that heritage. Can't get to the creamery? Highfield Farm Creamery's cheeses are woven into menus at Fire 2 Fork in Delavan, The Red Oak Restaurant in Bristol, and Savoy at the French Country Inn in Lake Geneva. They are also sold at Pearce's Farm Stand in Lake Geneva, and Laura's by the Lake, a modernized Lake Geneva liquor store with all the fixings for cocktails plus cheese, beer, wine, cutting boards, knives, beach-y apparel for women, coolers, and glassware. If you want to pull together an impromptu picnic of all your farm-food finds, this is your place.

County Fairs and Wisconsin State Fair

Jefferson County Fair
Jefferson County Fair Park, 503 N. Jackson Ave., Jefferson; 920-674-7148; jcfairpark.com
Spanning eighty acres, the fair pairs classic Wisconsin eats (cheese curds, Kiwanis baked potato, sweet corn, and Farm Bureau cream puffs) with live music at the grandstand and Miller Lite tent. Many animal species are shown at the fair, including swine, goats, cats, rabbits, poultry, and cattle. Air Max by Mr. Ed's Carnival is the most popular ride, according to fair organizers, and not for the faint of heart or those afraid of heights. Held: early July

Kenosha County Fair
Kenosha County Fairgrounds, 30820 111th St., Wilmot; 262-862-6121; kenoshacofair.com
Hosted in Wilmot since 1920, the fair's activities include a hay-bale throw contest, circus variety show, chainsaw carver Dave Watson

Kenosha Area Convention & Visitors Bureau

(makes art out of butter, chocolate, cheese, and ice), antique-tractor parade, petting zoo, pie auction, truck and tractor pull, and strolling musical entertainers (a new addition). There is also the crowning of the Fairest of the Fair (fair queen the following year). In addition to typical fair judging, food, and live music, there is the hilarious lamp-decorating contest. When: late August

Racine County Fair

Racine County Fairgrounds, 19805 Durand Ave., Racine; 262-878-3895; racinecountyfair.com

Whether it is sipping Wisconsin craft beers and wines in the beer garden, listening to polka, admiring antique tractors, watching a livestock auction, or witnessing the Fairest of the Fair crowning (plus Little King and Little Queen, for kids), there is something for everyone, in addition to opportunities to meet and greet with animals as well as hop on amusement-park rides. Knitting, crocheting, baking, woodworking, and other skills—along with animal livestock—are also judged. Foods include cream puffs, corn dogs, and deep-fried cheese curds. When: late July-early August

Walworth County Fair

Walworth County Fairgrounds, 411 E. Court St., Elkhorn; 262-723-3228; walworthcountyfair.com

At the state's largest county fair (since 1949), you will find edible delights like Knights of Columbus cream puffs and Wurst Haus brats, plus live bands (and Walworth County Idol shows), and barns dedicated to animal species. Livestock shows and auctions are also hosted and if it is too hot or raining, duck into the Arts and Crafts Building or buildings filled with antiques and horticulture. Kids love the carnival rides and Barnyard Adventure, with fun tutorials in farm life, plus a butterfly barn, composting, and rainwater recycling. When: Wednesday before Labor Day to Labor Day.

Washington County Fair

Washington County Fair Park & Conference Center, 3000 County Hwy. PV, West Bend; 262-677-5060; wcfairpark.com

This fair has been in operation since 1858—ten years after Wisconsin became a state—and attracts headliner musicians with big-name appeal (like rock band 38 Special, country-music star Brantley Gilbert, and American Idol winner Scotty McCreery). With deep-fried Oreos to cotton candy—and savory eats like walking tacos and sirloin-tip meals—nobody leaves hungry. Animal viewing includes swine, goats, cows, and chickens. Locals' arts and crafts, plus baking, skills are judged, too. Held: late July

Waukesha County Fair

Waukesha County Exposition Grounds, 1000 Northview Rd., Waukesha; 262-544-5922; waukeshacountyfair.com

Arrive with an empty stomach, as food ranges from Red Velvet funnel cakes to deep-fried olives on a stick. Cover bands perform daily and other activities include eating contests, Little Farmers Sheep Rodeo, Friday-night Jr. Livestock Auction, tractor pulls, carnival

rides, animal exhibits in five barns and tents (from teeny rabbits to enormous cattle), community members' woodworking and other projects, and 4-H Youth events. Do not miss pig, goat, and duck races. Sweet-tooth alert: cream puffs sold at the 4-H Kitchen. When: Late July

Wisconsin State Fair

Wisconsin State Fair Park, 640 S. 84Oth St., West Allis; 414-266-7007

By far the state's largest state fair—and among the country's largest, too—this ten-day-long celebration of farming life is hosted at State Fair Park in West Allis, just seven miles from towering skyscrapers in downtown Milwaukee. Due to the urban density, many choose to park their vehicles on locals' lawns for a small fee, in lieu of a parking lot. Each year the fair attracts about 1.1 million people—about double the size of Milwaukee. Since 1892 it is been held at the present site.

Food ranks high here, with Original Cream Puffs's cream puffs doled out at a dedicated building and aisles upon aisles of fried, decadent fair foods. A slide stretching several stories is practically an annual highlight for many fair visitors. During the mid-1990s, the Wisconsin Wine Garden was added, highlighting wines made in the state using cold-hardy grape varietals. All are available for sipping by the glass. Like other fairs, farm animals, arts and crafts, and culinary arts are exhibited in dedicated buildings. There are also quirky events like goat and pig races for some giggles. Hosted in the milking parlor are demos on how to milk cows and goats. You can also sip a glass of fresh, ice-cold milk for $.50 in signature flavors that change each year. In 2019 they were chocolate peanut butter, green mint, strawberry, chocolate, and root beer.

Lots of kid-friendly options are at the fair, including We Energies Energy Park, teaching about safety surrounding electricity and gas, as well as a solar-car race, and cooking and gardening demos. Each year the food line-up changes and is a contest of "the weirdest and best." Carnival rides send fairgoers squealing and screeching into the night. Big-name entertainers with national appeal—The Beach Boys, Brothers Osborne, and Boyz II Men—perform on the Main Stage nightly. Held: 11 days in early- to mid-August

Dairy Centers

Cheese Counter & Dairy Heritage Center
133 E. Mill St., Plymouth; 920-892-2012;
cheesecapitaloftheworld.com

For a quick primer in Wisconsin-cheese history, drop by this four-year-old educational center with interactive touchscreens. Establishing in Plymouth is intentional: this town is the state's cheese capital, home to producers Sargento and Sartori, and 15 percent of the nation's cheese is made right here. Score a "Keep Calm and Have Cheese" T-shirt along with Wisconsin cheese in the shop. Break for lunch at the Cheese Counter where all fourteen sandwiches and three mac n' cheese recipes fold in cheese. For dessert? Kelley Country Creamery ice cream.

Farm Museums

Old World Wisconsin
W372 S9727 Hwy. 67, Eagle; 262-594-6301; oldworldwisconsin
.wisconsinhistory.org

Any kid who grew up in Wisconsin likely took a field trip to this living-history museum that is an interactive step into the past to learn how the state's immigrant farmers functioned during the nineteenth century and into the early twentieth century. Finnish, Danish, Yankee, Norwegian, Polish, Pomeranian, and Hessian farmsteads aim to replicate a day-in-the-life during the 1860s to 1880s through arts and crafts, gardening, blacksmithing, animal husbandry, harvesting vegetables, and more. A tram enables accessibility for everyone.

Farms and Farm Stands

Alice's Garden Urban Farm
2136 N. 21st St., Milwaukee; 414-687-0122; alicesgardenmke.com

After a visit to this urban garden and farm (in Lindsay Heights, just down the road from The Tandem, a restaurant aided by José Andrés World Central Kitchen non-profit; the pay-what-you-can Tricklebee

Cafe; and the year-round Fondy Food Center), you will never view a flower or plant the same way again. Farm ecologist Sierra Taliaferro refers to every flower, vegetable plant, and medicinal herb as a "relative" and encourages asking the plant for permission before snipping or cutting its leaves. On guided biodiversity walks she dispenses advice for how to cook with what is grown here, whether it is borage brewed into a diuretic tea or amaranth woven into a Rice Krispie–like treat.

Located on the site of a former underground railroad, the garden debuted in 1972, under the care of Venice Williams, who continues to be active today. "When we cultivate the land, we're cultivating the history," says Taliaferro. "When you look at those flowers, remember that resilience and strength. Think of that power within yourselves when you view flowers."

Open for both self-guided visits and organized events—from yoga (including dedicated classes for kids) and meditation to vegan-food pop-up dinners, and also guided walks on the "living labyrinth," artisan markets, and farmers markets—Alice's Garden Urban Farm also houses community-garden plots for locals. Want to get your hands dirty? Volunteer work days are organized each season, too. During the COVID-19 pandemic and subsequent civil-rights battles last year, Williams organized Safe Space Sundays for members of the black community to come together, both to grieve and be empowered.

Amy's Acre

8318 6 Mile Road, Caledonia; 414-323-2210; amysacre.com
Owned by Amy Wallner, with help from her partner Jay, two cats, and a flock of hens, the bounty includes vegetables (known for their heirloom tomatoes), eggs and—soon—hops. Shop Amy's farm-fresh produce at either the twenty-acre farm's stand or at the Fox Point Farmers Market and Shorewood Farmers Market (both on Sundays), and you may also find it on a restaurant menu in the Milwaukee area. Pasta Tree, Ca'Lucchenzo, Morel, Miss Molly's Cafe, and La Merenda are huge supporters.

Bower's Produce
W490 WI-20, East Troy; 262-642-5244

Just three miles off I-43 is this produce stand/greenhouse grower touting succulents, culinary herbs, ornamental grasses and annual flowers (such as marigolds and hanging flower baskets) sold individually or by the flat. There are thirteen greenhouses on site. Produce is in perfect pitch with the season (late July through late November), whether that's corn, sweet peas, sugar-snap peas, black raspberries, tomatoes, peppers, kohlrabi, or beets. Check out the towering, antique barn with its weathered exterior and a painted American flag wrapped around it.

Cozy Nook Farm
S11 W30780 Summit Ave., Waukesha; 262-968-2573;
cozynookfarms.com

The Wendts founded this working dairy farm in 1834 at what's now Goerke's Corner. In 1957, when I-94 was being built, the operation moved to a new site. Current owners, since 1985, are Joan (Wendt) Oberhaus, her husband Tom, and their son Charlie. During autumn, dozens of varieties of pumpkins, gourds, decorative corn, and squash are sold—and hayrides can be booked on weekdays—followed by holiday wreaths, trees, and garland (grown in Northern Wisconsin) beginning the Friday after Thanksgiving. Between spring and fall, farm tours geared for children are by appointment.

Ela Orchard
31308 Washington Ave., Rochester; 262-534-2545; elaorchard.com

Co-owned by *New York Times* best-selling author Jane Hamilton and her husband Bob Willard, this third-generation apple and pear orchard retails through its Apple Barn. Otherwise, find them (along with apple cider) at the Greendale Downtown Market, Shorewood Farmers Market, and Dane County Farmers Market. Lodi, Jonathan, Melba, Red Delicious, Honeycrisp,and Black Willow Twig are among the twenty-six apple varieties, along with four varieties of pears (including Bosc and Moonglow). Yarn skeins spun from the orchard's sheep fibers are also sold.

Elegant Farmer

1545 Main St., Mukwonago, 262-363-6770; elegantfarmer.com
If Dean & DeLuca had a rustic, Midwestern cousin, this Mukwonago
farm stand would be just that. Look for the yellow smiley face atop
the towering barn along Main Street. (Dig the smiley face? Snap up
a T-shirt or frosted cookie as a souvenir.) It is entirely indoors and
open year-round.

A quick swoop through the store—just a few minutes off I-43 S/
US-45 S, thirty miles southwest of downtown Milwaukee—via the barn's
silo and you can load up a cart with a week's worth of groceries. Every
single item is artisan-produced and nearly all are made in Wisconsin.

From staples iconic to Wisconsin—such as blocks of cheese
(plus cheese curds) from award-winning creameries, summer sausages and brats—as well as bagged treats like popcorn in flavors that
go above and beyond caramel and cheddar cheese, not to mention
chicken or turkey pot pies you just pop into the oven, the focus is all
about celebrating local. Bottles of maple syrup; cans of preserved
fruits and pickled vegetables; and jars of applesauce and salsa also
line shelves. There is even a meat deli, grab-n-go salads, The Chocolate Factory's ice cream by the scoop, and cider-brined ham, folding
in the farm stand's specialty: apples. Frozen Door County cherries
and Michigan blueberries can be found here as well, with no need to
drive to where they are grown. When in season, fresh strawberries,
apples, and pumpkins are available.

The Elegant Farmers crown jewel is its Apple Pie Baked in
a Paper Bag®—also available in pineapple, Door County cherry,
rhubarb/apple, strawberry/rhubarb, blueberry/apple, peach/apple,
and caramel/apple. This is also a popular mail-order item through
the farm stand's robust catalogue business that includes gift baskets.
Found in the back of the store, you may consider picking up snacks
for the car or tomorrow's breakfast, whether it is a muffin, fruit turn-
over, or frosted sugar cookie.

Since 1970, The Elegant Farmer has been in its current location
but the business actually began in 1946 as a family-owned dairy farm
in Mukwonago operated by Dave and Elmer Scheel. A farm stand on
the property sold fruit. Today the owners are longtime employees
John Bauer and Keith Schmidt, plus John's brother Mike Bauer.

Before you hop back into the car, peek across the street: the orchard supplies much of the fruit for the market's pies and pastries and hosts pick-your-own apples, pumpkins, and squash. Picnic tables under an awning are a great place to dive right into your carefully chosen edibles if you are not ready to depart just yet.

Geneva Lakes Produce Greenhouse & CSA

W2083 WI-11, Burlington; 262-206-1271; genevalakesproduce.com

When resort-goers in the tony town of Lake Geneva want to visit a farm stand, they are often directed to the super-cute Geneva Lakes Produce Greenhouse & CSA. It is only a fifteen-minute drive away and owned by sixth-generation farmers (the Kosters) who believe in farming sustainably. Visitors are greeted by a greenhouse on one side, produce on the other (selling everything from locally grown tomatoes to corn, plus honey and River Valley Ranch spreads), with vintage wheelbarrows housing mini floral gardens.

Godsell Farm

S105W15585 Loomis Dr., Muskego; 414-425-2937; godsellfarm.com

Mark and Pam Godsell have farmed nine acres since 1990, servicing CSA customers with meat and vegetable shares, and opening to the public. Kids can host a birthday party; adults take classes in gardening, soap making, bread baking, crafting cheese and pasta, and backyard chicken raising; and tours are by appointment. The Godsells bring the farm to local classrooms via its hatchery program. Pick up eggs and meat (sausage, or the Meat Eater Basket with New York strip, pork shoulder, and beef chuck roast) at their roadside stand.

Good Oak Farmstead

N4647 County Road O N, Delavan; 262-215-1224

Five years ago, Rebecca and Nathan Barth fled corporate jobs in Chicago—"to slow down life and raise our family," says Rebecca—to transform a five-acre parcel of a former dairy farm. You can literally stock your kitchen with much of what they grow, forage, bake, and harvest, all sold at their farm cart, including vegetables, jams, and jellies (from their orchard's fruits), pre-ordered breads for Sunday pick-up (garden focaccia is a customer favorite), grass-fed beef (from their Dexter cattle), and chicken and duck eggs.

Grassway Organics
W2716 Friemoth Rd., East Troy; 920-894-4201; grasswayorganics.com

With live music and wood-fired pizzas, pizza nights are on Friday and Saturday between May and October. Cheese and meat toppings are grown on site, veggies sourced from a local farm, and ancient grains grown in nearby Elkhorn. A "health food" store sells organic bulk items; milk, eggs, and cheese; grass-fed beef; pasteurized chicken, turkey, and pork meat. Owners (since 2016) Chaz and Megan Self since 2016, the couple is raising their three sons in a setting wildly different from their own upbringings, with 35 Jersey cows grazing 210 certified-organic acres.

Gwenyn Hill Organic Farm and Gardens
N130 W294 Bryn Dr., Waukesha; gwenynhillfarm.com

Spanning 450 acres in Delafield—twenty-five miles west of down-town Milwaukee, and six miles from Waukesha's downtown—this farm originally settled by Welsh immigrants in 1842 embarked on a new chapter with its 2018 debut. Gwenyn, which means "honeybee" in the Welsh language, seeks to honor these roots.

After five successive generations of the Williams family farmed here, and they decided to let go of the land, the farm snagged approval to subdivide into 140 lots. Thankfully, this new-home con-struction nightmare was rescued by a buyer with a different, more pastoral vision, "keeping this beautiful tract in farming for years and years to come," says farm manager Linda Halley.

"As with most farms in southeast Wisconsin, the farm was originally a dairy farm with sheep, as well. Our dedication to grow-ing food crops, rather than commodity crops, has led us to growing certified-organic vegetables, grains, fruits, grass-fed livestock, and organic eggs," she says. Recently the farm reintroduced dairy, selling certified-organic milk to Westby Co-Op Creamery in Westby. Also grown on eleven acres are certified-organic vegetables, berries, grains, hay, and flowers, joining certified organic eggs, grass fed lamb and beef, and Meadow veal. Although not certified-organic, honey and maple syrup are crafted using sustainable methods. Lloyd Williams—who grew up on the land—serves as crops contractor and also makes Bon Bree brick cheese.

Customers who receive a weekly CSA box (opting for shares in meat, vegetables, or eggs) also get access to the you-pick garden. The farm also staffs a stand at the Brookfield Farmers Market. Branded merchandise like T-shirts and stainless-steel water bottles help spread the word about one of Waukesha County's newest farms.

On a visit, you can shop at the farm stand—selling dairy products, vegetables, meat, and eggs—but to go deeper it is encouraged to sign up for a class in a topic such as organic farming or floral design.

Like any small-farm business, especially one that's only three years old, ingenuity never wavers. New ideas are constantly tossed around, with an eye on attracting people to the products but also the farm lifestyle. "We are searching for custom milling services so we can offer flours and oats," says Halley, "as retail products in the near future."

Healthy Harvest Farm
N64 W18768 Mill Rd.; Menomonee Falls; 732-742-8942; healthyharvestfarm.org

Functioning as a non-profit, the farm provides some of its organically grown vegetables, fruit, and herbs—grown only from heirloom organic seeds—to local food pantries through Feeding America. Last year, the farm (run by Wisconsin native Kathy Hoffman, who returned home from New Jersey in 2015) launched its CSA program, and added fruit trees, strawberry bushes, and asparagus plants, and will continue teaching classes. Want to give organic gardening and farming a try? Volunteers are always needed during the week.

Holy Hill Art Farm
4958 Hwy. 167 (Holy Hill Road), Hubertus; 262-224-6153; holyhillartfarm.com

Blending art and live music with agriculture, this 160-year-old, eighty-acre farmstead (locals know it as Loosen Family Farmstead), is open select weekends between June and mid-October. Every weekend is a different theme, with past examples like a Door County fish boil, Tom Petty tribute band, stone-fired pizza and—in September and October—the beloved Art & Farm Market. Dine in the historic barn for a ticketed, all-inclusive dinner event, seated at farm tables adorned with flowers cut from the gardens, and paired with live music.

Jerry Smith Produce and Pumpkin Farm

7150 18th St. (Highway L), Kenosha; 262-221-9765; jerrysmithfarm.com

Come autumn, this farm—midway between Milwaukee and Chicago—is an elbow-bumping situation on weekends because it is that much fun, with hand-painted pumpkin displays, a corn maze, pony and camel rides, a petting zoo, hayrides, and a Country Store. Other times of the year are pick-your-own days (vegetables and fruits), and weekly Crop Boxes filled with produce are a huge hit. Green Acres—a farm stand at the corner of 60th Street and Highway 31 (Green Bay Road) in Kenosha—sells nursery-grade plants, honey, apple cider, and vegetables.

Jelli's Market

N5648 S. Farmington Rd., Helenville; 262-593-5133;
jellismarket.com

With stands in Johnson Creek, Watertown, and Wales, residents
of Walworth and Jefferson Counties are never far from sustainably
grown produce (sweet corn, asparagus, apples, and peaches) along
with meat (Angus beef, lamb, pork, turkey, and chicken), soap, honey,
and frozen pies. A commercial kitchen added in 2017 means jams,
pies, and apple-cider donuts. You-pick days for blueberries, raspberries, and green beans in June and July are followed by a vibrant
sunflower field come late July. Green thumbs love the plants and
flowers sold at the greenhouse.

Jones Dairy Farm

601 & 800 Jones Ave., Fort Atkinson; 800-635-6637;
shopjonesmarket.com and jonesdairyfarm.com

While this breakfast ausage and Canadian bacon maker on a three-
hundred-acre farm could easily be confused with Bob Evans (the
packaging is that slick!), since 1889 it has been in the Jones family.
Milo Jones' great-great-grandson is the current CEO. A vintage
post-and-beam barn, Colonial Revival home, and more are on the
National Register of Historic Places. Take a look for yourself at Jones
Market next door, in a 1906 building selling meats and Cedar Crest
hand-dipped cones and ice-cream floats. Tours on select dates.

Larryville Gardens

W1349 WI-11, Burlington; 262-206-2360; larryvillegardens.com

This is not just any four-acre vegetable farm. Owner Larry Cannon
and his wife Michelle specialize in growing organic ingredients, like
heirloom tomatoes, that an Italian chef may whip up into a meal.
They are a "teaching farm" in that they school homesteaders in
canning, making tomato sauce, and cooking with fresh produce. An
annual farm-to-table dinner, wood-fired Pizza Nights (with Tomato
Pie Co.), and Wednesday pop-up farmers market are joined by a
farm store (partnering with makers of goat's-milk soap, grass-fed
meats, and sourdough bread) born out of a packing shed.

LotfotL Community Farm

W7036 Quinney Rd., Elkhorn; 920-318-3800; lotfotl.com

It is a question couple Tim Huth (the farmer) and April Yuds (handles marketing, the CSA, and customer service) get asked often: What does LotFotL stand for? Answer: Living off the fat of the land. After Tim started the farm (harvesting pigs, turkeys, vegetables, fruits, and honey) at Michael Fields Agricultural Institute, he moved it to the Quinneys' historic dairy barn in 2011. An annual farm-to-table dinner is organized with Dave Swanson, chef/owner of Milwaukee's Braise. The farm also operates a stand at South Shore Farmers Market in Milwaukee.

Lucky Break Acres

W6022 County Road A, Elkhorn; luckybreakacres.com

In 2014, Anthony and Michelle DiMauro moved from Las Vegas into their 1830s farmhouse and former dairy barn. For their weekends-only farm stand, they craft beeswax candles and lip balm, watercolor notecards, cut flowers, and strawberry-jalapeño jam; and also harvest free-range eggs, honey, vegetables, and herbs. "We had never farmed and wanted to try a completely different life," says Michelle. "My husband had family who lived in Salem when he was growing up in the Chicago suburbs. He used to spend holidays, weekends, and

summers there and it made him fall in love with Wisconsin." Yoga classes are paired with floral workshops and a tour.

Michael Fields Agricultural Institute

N8030 Townline Rd., East Troy; 262-642-3303; michaelfields.org

Every entrepreneur needs a boost. Farmers just starting out can get it at Michael Fields Agricultural Institute, which operates with staff and a board of directors, and is one of the nation's leading advocacy groups for sustainable farming.

The beginnings of several farms around the state can be traced back to its intern and apprentice program, which acts as an incubator, with the idea that the farm will continue to profit in another location. Alumni include a former garden intern who left an engineering career to open a bakery in France, on his family's farm. And a Botswana native, after the institute's training, is now a lead farm educator at City Growers in New York City. One couple (she is from Belgium, he is from Green Bay) even met as apprentices at the institute. What is unique is that the focus is not just on one type of crop. Instead, it covers many, so long as there is an intent to build a career out of farming, whether it is planting (and culling bouquets from) a flower field or harvesting vegetables to sell to chefs.

The institute—founded by Christopher and Martina Mann, and Ruth Zinniker in 1984—has tirelessly advocated for continued support of sustainable-agricultural research, education, and policy work. It keeps an eye on not just rural farms but urban farms, too, especially those supplying farm-fresh produce to so-called "food deserts" (pockets in a city that lack access to healthy, nutritious food). One example of lobbying for all these mantras includes creating, funding, and implementing Wisconsin's Buy Local, Buy Wisconsin initiative. Another example: securing grant funds for the Iowa County Uplands Farmer-led Watershed Group, designed to improve water quality flowing into the Gulf of Mexico. An organic-corn breeding program is in partnership with the USDA.

Some of the agricultural practices the institute supports include cover crops (which protect and enhance soil while also reducing erosion by capturing raindrops), biodynamic farming, and no-till organic farming,

Keep an eye on the institute's website for news of conferences, field days, and workshops. Small groups can also book a tour.

Morning Star Family Farm
4504 State Road 83, Hartford; 262-670-6561; morningstarfamilyfarm.net

Is your kid super into farming and want to learn more? The Roxas organize Farmer for a Day camps at their certified naturally grown farm each summer. Practices in raising vegetables and learning what makes animals healthy are taught. The farm also invites you out for you-pick fruits (raspberries, Asian pears, plums, and apples) by appointment; brews and sells kombucha; and offers Meat CSA subscriptions. Pre-orders (not just vegetables and meat but also elderberry syrup and bone broth) are delivered to areas around Milwaukee.

Nourish Farms
100 Alfred Miley Ave., Sheboygan; 920-550-2020; nourishfarms.org

From three-course dinners to events in the gorgeously restored Miley Barn (with Edison lightbulbs and a lofted ceiling), as well as farm-to-table tours and cooking-class dinners (for private groups of up to eight people), people come here for many reasons. All are about reconnecting with food and the soil, tying into the mission of helping to create a healthy community through food security. The farm was founded in 2009. Unlike many farms, which are often boot-strap operations, this one relies upon full-time staff but also volunteers who do not mind getting dirty.

Pearce's Farm Stand
W5740 N. Walworth Rd., Walworth; 262-575-3783; pearcefarms.com

Since 1955, the Pearces have operated this farm, only the second family to do so since 1848. Fall is when the farm really comes alive—offering adventure for the kids (such as a corn maze, hayrides, and tractor-pulled wagon rides) and locally made foods (honey, jams, jellies, donuts, and kettle corn; plus produce like sweet corn grown

© Holly Leitner

on the farm) for all ages. Pumpkins for carving and gourds for decorating are also sold. A car show is in late August.

Quednow's Heirloom Apple Orchard

W5098 County Road ES, Abells Corners; 262-501-9033; quednowsappleorchard.com

This family-owned orchard is known for options: about ninety different varieties are grown, which translates to ten types available at once. These are sold at farmers markets in Fort Atkinson, Whitewater, and Whitefish Bay (look for its Apple Cider Muffin Tin donuts in Fort Atkinson along with a family member's Jazzed-Up Marshmallows). During the season (August to late October) the orchard is open Friday, Saturday, and Sunday. Integrated pest management philosophies guide the orchard, which also produces pears, peaches, and cherries (sweet and tart).

River Valley Ranch & Kitchen

39900 60th St., Burlington; 262-235-2555; rvrvalley.com

If you have spent time in Chicago's Lincoln Square neighborhood, this business may be familiar (there is a shop and eatery called River

Valley Farmers Table there). For years, I had driven past the Burlington farm store, tucked into a barn along Highway 50, until a foodie friend clued me in. Inside, you will find lunches in the cafe (such as tamales topped with portabella), grocery-like goods that include exotic mushrooms and the brand's canned products (like portabella pasta sauce), and outside is a perennial favorite for home gardeners: mushroom compost.

Rushing Waters Fisheries & Trout Farm
N301 County Road H, Palmyra; 262-495-2089; rushingwaters.net
Pick a restaurant in the Milwaukee and Lake Geneva area serving rainbow trout and chances are it comes from this eighty-acre farm—open to the public for pond fishing. (They will even fillet it for you.) Not ready to pick up a pole? Its farm store sells rainbow trout (smoked or fresh) along with Wisconsin wines, beers, condiments, and snacks—plus wild-salmon burgers and smoked-salmon spread. Or try the trout for yourself in the restaurant. Field biologist Peter Fritsch founded the fisheries in 1994 with Bill Graham but the property's trout-farming history dates to the 1940s.

Schmit's Farm Produce
10333 N. Wauwatosa Rd., Mequon; 262-242-3330
Known for its sweet corn come July, this farm stand also brings its bounty of vegetables (such as green beans, new potatoes, tomatoes, zucchini, and Delicata squash) to farmers markets in Greendale, Grafton, and Thiensville in case the stands' locations are not convenient to you. Pumpkins arrive in the fall.

Soap of the Earth
N221 Coldspring Rd., Whitewater; 608-359-1015;
soap-of-the-earth.com
This farm with an 1872 farmhouse is different in that the bounty is bars of soap, deodorant, balms, facial serums, insect repellant and shower gel—not bundles of vegetables. But milk from Nubian goats—who owner Lori Hoyt calls "business partners"—and essential oils plus oils (olive, coconut, and palm) are woven into each soap batch, sold at the on-farm stand as well as at the Whitewater City Market,

Dane County Farmers Market, Mileager's Great Lakes Farmers Market, and Milwaukee Winter Farmers Market.

Stiles Vegetable Farm & Greenhouse
11717 Sheridan Rd., Pleasant Prairie; 262-694-5256
Since 1950, this family-owned farm stand has been a pit stop (and, by now, childhood memory) for many locals, just north of the Illinois state line in Kenosha County. Fruits and vegetables grown at the farm are sold here, along with plants, herbs, and flowers, plus Amish canned goods. A favorite find are ears of corn and tomatoes from Georgia and Southern Illinois that arrive each June and July, extending the season.

Stone Bank Farm Market
N68 W33208 County Road K, Stone Bank; 262-563-8010; stonebankfarmmarket.com
Rocking a modern-farmhouse aesthetic (you practically expect to see "Magnolia" queen Johanna Gaines come out to greet you!), this working farm's year-round market is in tiny Stone Bank, on the same plot of land as the farm. Tucked into an 1858 darling Presbyterian church with fresh coats of white paint both in and out, it is open Wednesday through Sunday for a few hours each day.

Here, you can pick up all the goods for a nearly fully stocked kitchen, including dairy products, meat (poultry, lamb, organic pork, and organic grass-fed beef), honey, yogurt, maple syrup, cheese, and organically grown vegetables that run the gamut from heirloom tomatoes to lettuce leaves. Most of it is grown or produced on the seventy-five-acre farm, which includes not only the market, but also a three-acre vegetable garden and a four-thousand-square-foot greenhouse. The farm's roots date back to 1844 when a Scottish immigrant scooped up the land and developed Stone Bank, using this plot for farming. It's currently owned by the Faye Gehl Conservation Foundation. Other items for sale in the market are procured from regional growers and farmers that include Olden Organics, Waseda Farms (a certified-organic meat producer in Door County), Turtle Creek Gardens, and Zaiger's Clover Meadows (a dairy farm in Marathon producing A2 milk from cows that feed on grass and are free of antibiotics and growth-hormones).

If you are out and about on a little road trip, consider dropping in for a hand pie, handful of cookies, or tart—for now or later. It is likely that seasonal fruits are baked into your selection.

If you want to go deeper—and perhaps become a farmer yourself—the market hosts workshops and classes in its education center that cover gardening, cooking, and nutrition. Kids can learn and play with a licensed teacher who leads two- to three-day sessions throughout the summer. Stay tuned to the website for news of its walking farm tours that—with an eye on fostering appreciation for all the work it takes to get farm-fresh food into consumers' hands—take guests to the beehives as well as on a visit with resident cows, pigs, and chickens. Saturday-morning yoga classes in the barn and occasional Wednesday pizza nights also draw people to the farm. Each spring there is a plant sale and, in September, the curated Homegrown Market with vendors, live music, tractor guides, organic food, and farm tours, followed by the annual Vintage Xmas pop-up décor sale.

The Healing Place Farm
N87W22211 N. Lisbon Rd., Sussex; 262-370-3810; thehealingplacefarm.com

Farming is not just about vegetables—it is also about beauty and self-care. Founder Polly J. Huenink-Schellinger (an aromatherapist and nurse) teaches classes in aromatherapy as well as "make and take labs" at her family's farm that are open to the public, as well as private groups of five or more. Sold at the farm store are bars of goat's-milk soap plus aromatherapy sticks and rollers, sprays, blends, balms, salves, body butters, and a men's line, along with a protectant for a dog's paws and nose.

Three Brothers Farm
N87N34109 Mapleton Rd., Oconomowoc; 262-470-4429; threebrothersfarmcsa.org

Michael and Courtney Gutschenritter bought his grandparents' one-hundred-acre farm in 2017, becoming the family's first generation to make a living as farmers. Inspired by working on farms in Maine and Italy—and with a painting degree—Courtney grows flowers and teaches workshops through Courtney Joy Floral. Their pasture-raised, organically fed eggs go to Hunger Task Force, the Piggly Wiggly in Oconomowoc, and restaurants like I.D. in nearby Delafield. They also sell wool and raise and sell grass-fed lamb and beef. Concerts with pizzas, as well as tours, bring people to the farm.

Witte's Vegetable Market, LLC
2313 Highway NN, West Bend; 262-338-4589; wittesvegmarket.com
Two couples in the Witte family operate this market, born out of Dave Witte's nostalgia growing up on a farm as one of eight kids. In

© Witte's Vegetable Market, LLC

2010 the market opened at the farm, which has since expanded to 105 acres, of which 60 are farmed using integrated pest management practices. Pick-your-own tomatoes in late summer are joined by dozens of vegetable crops (from asparagus to Yukon potatoes) for sale, plus Michigan blueberries, local maple syrup and honey, and Barthel apples.

Zinniker Family Farm
N7399 Bowers Rd., Elkhorn; 262-642-5757; zinnikerfarm.com
This biodynamic farm—in operation since 1943—is the country's oldest biodynamic farm. Currently managed by third-generation owners Petra and Mark Zinniker as a beef operation, they host workshops and events to generate excitement about farming. This includes Spring Prep Day and Fall Prep Day, windows into the launch of a farm's season. Soil and compost research are cutting edge, influenced by Sanskrit recipes in India as well as Korean natural-farming methods. Beef, soap, and "biodynamic preps" (as a spray or added to compost) are sold through the farm.

Farm Stays

Abloom Farm Resort
2839 State Road 33, Saukville; 262-387-1001; abloomfarm.com
Owners Dale Stenbroten and Katy Rowe offer up a six-bedroom, newly renovated retreat on their seventeen-acre farm perfect for multiple generations of the same family or perhaps a girlfriends' getaway. It is only thirty minutes north of Milwaukee and tucked into a pine forest, with a fire pit and hiking trails, plus a cute log cabin and two barns. Tours are by appointment and the venue is also rented out for weddings (it is that stunning!) and retreats, with the farm's wellness coordinator assisting for spa and private-chefs services, plus yoga.

The Country Retreat at The Historic Schmitz Family Farm
W188N12369 Maple Rd., Richfield; airbnb.com/rooms/16200052
Bookable through Airbnb, this two-bedroom, four-bath home is snug on the Schmitz family's hobby farm in Germantown where not only

do chickens scurry around but so do Laura and Matt's (the owners) two Goldendoodles. Since 2015 they have rented it out. Whether you plan to cozy up to the fireplace, hike on the Ice Age Trail or in the nearby Kettle Moraine National Forest's Northern Unit, or view a stunning sunset from the deck, the goal is to introduce you to a night on the farm. On-farm massages available through a local therapist.

The Inn at Paradise Farm

3894 Paradise Dr., West Bend; 414-218-8558; innatparadisefarm.com

Jim and Mary O'Connell's blissful pad open to bookings is their family's 1847 four-room log cabin on ten acres, with a wood-burning stove, outhouse, kitchenette and, if you desire, a glamping tent that can sleep up to six people (it is that big!). Guests are welcome to harvest vegetables when in season and procure fresh eggs from their free-range chickens. Want to go deeper? Private homesteading workshops can be arranged.

Farmers Markets

Brookfield Farmers Market

Brookfield Central High School's front parking lot, 16900 W. Gebhardt Rd., Brookfield; 262-784-7804; brookfield farmersmarket.com

This market with around one hundred vendors dates to 1991. Stands sell items like Door County cherries and peaches, Italian sausages, microgreens, vegan popsicles, elk meat, and Harmony Specialty Dairy cheese. The third Saturday of the month adds arts-and-crafts vendors. Shopping for someone's birthday? You will find fun and unique items, such

© Brookfield Farmers Market

as retro hand-sewn aprons or barnwood crafts. You can even pick up worm castings for your garden or have knives sharpened while you shop. Saturdays from 7:30 a.m. to noon, early May through late October.

Brown Deer Farmers Market

9078 N. Green Bay Rd. (by Firehouse Subs), Brown Deer; 414-371-3000; browndeerfarmersmarket.org

What this market does differently is stay open all day—not just a short window of time. Shoppers can pick up sacks of caramel corns, honey, canned or pickled vegetables, and vegetables and strawberries grown by Hmong farmers. Just-made crepes, hot-off-the-grill Junior's Smoked BBQ, Dessert Land's Thai cuisine, and live music allow for a little bit of lingering. Wednesdays from 9 a.m. to 5:30 p.m., mid-June to late October.

Burlington Farmers Market

355 N. Pine St. (Wehmhoff Square), Burlington; 262-210-6360; burlingtonwifarmersmarket.com

Livening up downtown Burlington every Thursday since 2008, vendors sell everything from cheese (Hill Valley Dairy), milk (Oberweis), and vegetables (Seeds of Hope, which provides opportunities to disabled adults is one vendor) to artisan breads, Hemken Honey Co. honey, and Soap of the Earth farm-made soaps. Roasted coffee beans and Greek olive oil from food artisans are joined by prepared Brazilian and Mexican cuisine, allowing shoppers to linger. Thursdays from 3 to 7 p.m., May through October.

Cathedral Square Market

Cathedral Square Park, 920 E. Wells St., Milwaukee; 414-271-1416; easttown.com/cathedralsquaremarket

Each summer, locals flock to Cathedral Square in downtown Milwaukee's East Town neighborhood for two reasons: the Saturday farmers market and Thursday Jazz in the Park concerts. Operating as an incubator for small businesses and new vendors, the market offers only fresh vegetables and fruits, but also artisan-made crafts and jewelry, and to-go South American and Southeastern cuisine.

Tiny shoppers love The Nest Culinary School's free cooking classes. Saturdays from 9 a.m. to 12:30 p.m., mid-June through mid-October.

Delafield Farmers Market

Municipal parking lot at the corner of Dopkins and Main Streets, Delafield; 262-409-5276; delafieldfarmersmarket.com

Tucked into the darling Delafield—its downtown developed by Bob Lang in the late '80s, much like his stationary company's small-town scenes—this market sells heirloom produce, grass-fed meat, chicken and duck eggs, microgreens, honey, cheese, berries, mushrooms, maple syrup, popcorn, and more. There is even a gluten-free foods tent and you can pick up bars of locally made goat's-milk soap. Saturdays from 8 a.m. to 1 p.m., May through October.

Dousman Farmers Market

118 S. Main St. (across from fire station in Village Hall parking lot), Dousman; 262-968-4566; dousmanchamber.org

There is nothing you cannot find at this market. Eggs, meat, vegetables (from beets to sweet corn), fruits, plants, herbs, and even trees are sold here. Looking for a gift? Dog toys, macramé, bookmarks, wildlife-photo greeting cards, and kettle corn are a few options. Live music is also a part of the shopping experience. Wednesdays from 2 to 6 p.m., early May through mid-October.

The East Troy Market

Village Square Park, 2881 Main St., East Troy; 262-642-3770; Easttroy.market

East Troy may be a tiny town in Walworth County but it is surrounded by one of the state's highest concentrations of family-owned farms producing meat, fruits, and vegetables they bring to market (such as Rushing Waters Fisheries' salmon and Healey's Wholesome Haven's sweet corn). Each market is heavy on varied entertainment outside of the farmers market, including a food-truck fest, makers market, and artisan night.

Enderis Park Farmers Market

Enderis Park, 2956 N. 72nd St., Milwaukee; enderispark.org

This North Side neighborhood's market is relatively new (established in 2016 by the Enderis Park Neighborhood Association) and proof of the area's popularity for homeowners. A mix of local and regional vendors are there each week, including prepared foods from Miss Molly's Cafe and Pastry Shop, Wallah Mango Lassi, and Big Yellow Farmhouse Bakes, plus Happy Destiny Farms' chicken products and Ann and Jerry's off-beat produce (jicama and tomatillos, for example). Sundays from 9 a.m. to noon, mid-June through mid-September.

Fondy Food Center

2200 W. Fond du Lac Ave., Milwaukee; 414-933-8121; fondymarket.org

In 2019 a forty-acre farm debuted, just for the market's vendors. Ethnic roots drive shoppers' selections, including refugee Hmong farmers produce and Filipino noodles; and "Seasonal Soul" cooking demos where neighborhood chefs whip up family recipes, plus Saturday yoga. There is an annual BBQ fest in July. Saturdays from 9 a.m. to noon, early spring; Saturdays from 7 a.m. to 2 p.m., and Sundays, Tuesday, and Thursdays 9 a.m. to 2 p.m., early May through late November; Saturdays from 9 a.m. to 2 p.m., late season.

Fontana Farmers Market

441 Mill St., Fontana; 262-275-0040

Across from The Abbey Resort and in front of Coffee Mill, this farmers market attracts vendors like a local creamery (Highland Farm Creamery, which produces cheese curds and other products), vegetable and fruit growers, honey producers, and makers of soap and other crafts. Pop into Coffee Mill for a pastry and coffee or tea. Saturdays from 8 a.m. to noon, seasonal

Fox Point Farmers Market

7330 N. Santa Monica Blvd. (North Shore Congregational Church parking lot), Fox Point; fpfarmersmarket.com

Organic Red Stone Rice grown at the state's only rice farm, Madame Macaron MKEs cookies in a sea of pastels, Stamper Cheese's artisan *fromage*, Wallah Mango Lassi refreshing drinks, Elsie Mae's fruit pies, and Aleka's Kitchen's marinated Greek olives are among the delicacies sold at this farmers market, in addition to vegetables, eggs, free-range chicken meat, pasteurized beef, and smoked salmon, plus yummy, prepared eats like egg rolls, tamales and African cuisine. Grab a Wild Flour Bakery muffin to munch on as you roam. Saturdays from 8 a.m. to noon, third Saturday in June through third Saturday in October.

Garden District Market
Wilson Park (near the Wilson Park Senior Center),
1601 W. Howard Ave., Milwaukee
There are a lot of gardeners and urban farmers in this neighborhood, truly reflecting its namesake. Some have plots at the Garden District Community Garden. At the market you will find—in addition to locally grown fruits and vegetables—unique items like CBD oils, egg rolls, and original art by Milwaukee artists Susan M. Edgerton and Jamie S. Stoehr from Water's Edge Gallery and Retail Company. Saturdays from 1 to 4 p.m., late June through late October

Germantown Farmers Market
Germantown Village Hall, N112 W17001 Mequon Rd., Germantown;
262-250-4710
This market does not just sell produce, it also features natural-stone, artist-made jewelry; maple syrup; potted plants; cut flowers; farm-fresh eggs; lots of vegetables and fruits; and honey. Start your weekend off right by picking up these calming, good-for-you finds. Germantown was known as a farming community until 1963, so this market really goes back to the village's founding roots. Saturdays from 8 a.m. to noon, mid-May through late October.

Grafton Farmers Market
Twin City Plaza, 1720 Wisconsin Ave., Grafton; 262-377-1650;
grafton-wi.org
At this rare, day-long market in this North Shore community, organized by the Grafton Area Chamber of Commerce, you can score honey, flowers, fruits, and vegetables. Or—for a snack while you

shop—baked goods. Thursdays from 9 a.m. to 6 p.m., early July through late September.

Greendale Downtown Market
5602 Broad St., Greendale; 414-423-2790

This southwest Milwaukee suburb has an interesting design past: it is one of three Greenbelt towns established in 1936 (the others are in Maryland and Ohio). Come check out the community via its farmers market, in the historic downtown, with different vendors each week. Examples include Ela Orchard (apples), Yuppie Hill Poultry (eggs), Schmit Farms (corn and peaches), and Artas'n Meats (from bacon to summer sausage). Handmade crystal jewelry, woodcraft, and Bloody Mary mixes also sold. Saturdays from 8 a.m. to noon, June through early October.

Greenfield Farmers Market
Konkel Park, 5151 W. Layton Ave., Greenfield; 414-939-8329; greenfieldwifarmersmarket.com

At one of the region's largest farmers markets (about fifty vendors), listen to live music as you scoop up farm-grown, fresh-cut floral arrangements; jars of honey, jams, and preserves; eggs; Benhart Farms beef, swine, and chicken; pastries (including Happy Vegan, churros, and Mr. Dye's Pies) and—of course—seasonal produce. Samples from a local brewery and coffee roaster are a nod to local entrepreneurs as is the Makers Market on the first Sunday of the month. Sundays from 10 a.m. to 2 p.m., May through October.

Hartford Saturday Farmers Market
Hartford Recreation Center parking lot, 125 N. Rural St., Hartford; downtownhartfordwi.com

Sold here are all the staples of any good farmers market: cut flowers, organic eggs, beef, chicken, bakery items, honey, and seasonal fruits and vegetables. A Wednesday-night market—also organized by the Hartford Area Chamber of Commerce—is paired with live music and at the parking lot at Jack Russell Memorial Library, 100 Park Ave., Hartford. Saturdays from 7 a.m. to noon, mid-May through late October.

Horicon Farmers Market

Kiwanis Park, 760 S. Hubbard St., Horicon; 920-485-0216; horiconphoenix.com

Near one of the state's most treasured natural spaces—Horicon National Wildlife Refuge—this market requires vendors grow or make what they sell (no bananas sold here!). You will find crafts like catnip toys and fiber arts, along with caramels, hickory nuts, cheesecake, and pickled items, joining seasonal fruits and vegetables. Live music is scheduled 6 p.m. and 8 p.m. Wednesdays from 4 to 7 p.m., late May through early October. Moves indoors (Horicon American Legion Post 157, 735 S. Hubbard St., Horicon) come winter.

Jackson Park Farmers Market

Jackson Park, 3500 W. Forest Home Ave., Milwaukee

Vendors here rotate each week, reflecting what vegetables and fruits are in season and at their freshest. Find the market near the boathouse in this 113-acre Milwaukee County park. Prepared-foods options include eggrolls and sweet purple rice from Vang Sisters Hot Food and just-popped kettle corn, and you will also find honey, cheese, fresh flowers, and crafts, as well as Three Seeds Clinic's herbal teas. Thursdays from 3 to 6:30 p.m., mid-June to late September.

Kenosha Harbormarket

2nd Avenue between 54th and 56th Streets, Kenosha; 262-914-1252; Kenoshaharbormarket.com

From macarons baked by French nuns to Cheffrey's Fine Foods gluten-free chili-cheese fries, this market—founded in 2003—is among the state's best. Bonus: it is along Lake Michigan, with green space to plop down with your market finds. Arts and crafts are of high caliber, such as hand-crafted bubble wands and beach-glass jewelry (scored from the lake's beaches), and produce shifts with the seasons, including strawberries, sweet corn, pumpkins, and squash. Saturdays from 9 a.m. to 2 p.m., late June through late September.

Kewaskum Farmers Market

Village of Kewaskum parking lot, east of Highway 45 and Highway 28 intersection, Kewaskum

This small market has all the locally grown goods you would expect, such as ears of roasted sweet corn, mushrooms, cucumbers, tomatoes, carrots, and lettuce. You can also find chicken meat and eggs; house plants and cut flowers, plus offbeat finds, like quail eggs and wild-caught salmon from Alaska. A baker vendor sells dessert breads, fruit pies, cookies the size of your head, and dog treats. Thursdays from 2 to 7 p.m., early June through late October.

Lake Geneva Farmers Market

Horticultural Hall, 330 Broad St., Lake Geneva; 262-248-4382; horticulturalhall.com

Locals know that this resort town quiets down during the week—perfect for strolling through the farmers market, in an Arts & Crafts building hugging Geneva Lake and on the National Register of Historic Places. Vendors range from South Padre Seafood (Gulf of Mexico grouper and Alaskan sockeye salmon) to honey producers, plus tamales and empanadas makers, and lots of fruit and vegetable growers, too. Thursdays from 8 a.m. to 1 p.m.; early May through late October.

Milaeger's Great Lakes Farmers Market

Milaeger's, 4838 Douglas Ave., Racine; 262-210-6360; milaegers.com

Among items sold at Milaeger's Racine store's year-round market are the greenhouse's own microgreens as well as Yuppie Hill Poultry's eggs (in nearby Burlington), River Valley Ranch mushrooms, and Wisconsin Soup Co.'s soups. Properly celebrate Sunday at the Bloody Mary bar, also serving local beers, and paired with live music and baked goods from Lake Geneva's Simple Bakery. Sundays from 10 a.m. to 2 p.m., outside between May and September, then moves inside.

Milwaukee Public Market Outdoor Urban Market

400 N. Water St., Milwaukee; 414-336-1111;
milwaukeepublicmarket.org

Often compared to Seattle's Pike Place Market, this year-round hub
(all indoors) for food vendors and grocers—there is even a wine shop
with a wine bar (Thief Wine)—broadens its scope with the Outdoor
Urban Market. Growers, artists, and makers set up along St. Paul
Avenue. Saturdays from 10 a.m. to 2 p.m., early June through Labor
Day weekend. Looking for culinary inspiration, or tips on canning
and pickling? Cooking classes are held inside the market.

Milwaukee Winter Farmers Market

Mitchell Park Horticultural Conservancy's The Domes,
524 S. Layton Blvd., Milwaukee; 414-562-2282; mcwfm.org

Operated by the non-profit Fondy Food Center, whose farmers-market vendors shift here during winter, the market features about fifty vendors. Despite it not being an outdoor-farming season, there is a mix of breads and pastries (such as Mr. Dye's Pies), fresh eggs, cider, maple syrup, chocolate, soups, five cheesemakers, and pantry staples (from Oly's Oats to Red Stone Rice)—along with fruits and vegetables. Saturdays from 9 a.m. to 1 p.m., early November through late March (excluding holiday weekends).

New Berlin Farmers Market

New Berlin City Center, 15055 W. National Ave., New Berlin;
262-786-5280; newberlinchamber.org

After picking up farm-fresh produce—and jams, cinnamon rolls, and fried pies from Amish-owned Esther's Bakery, or maybe hot-off-the-grill breakfast sandwiches—you can drop into local shops and restaurants, including 1848 Coffee's cafe. Also sold at the market are flowers and starter plants for your own garden, along with kettle corn. Saturdays from 8 a.m. to noon, early May through late October.

Oak Creek Farmers Market

Drexel Town Square, 361 W. Town Square Way, Oak Creek;
visitoakcreek.com/farmers-market

Tucked into Oak Creek's newest "downtown" with restaurants, city buildings, apartments, and retailers, a market trip can be followed by lunch. From pasture-raised, grass-fed beef at one of the country's oldest biodynamic farms (Zinniker Family Farm in Elkhorn) to Brightonwoods Orchard's and Quednow Heirloom Apple Orchard's apples come fall, plus Oak Creek–grown Le Vang Orchard peaches, thirty vendors fold in flowers, honey, cheese, and pastries. Wiscotti's lavender-honey biscotti feature local lavender. Saturdays from 9 a.m. to 1 p.m., early June through late October.

Oconomowoc Farmers Markets

Bank Five Nine parking lot, 155 W. Wisconsin Ave., Oconomowoc (outdoors); and Oconomowoc High School, 641 E. Forest St., Oconomowoc (indoors); 262-567-2666; oconomowoc.org

What is great about this market is it does not quit come fall—instead, it moves to an indoor location. Lyna Cafe's bubble tea and Thai milk tea are a fun change from morning coffee as you roam among the vendors, who include Brew City Pickles, Pete's Pops (popsicles in fun flavors), Pine Hill Orchard (apples), and Elemental Bread Company (artisan-made sourdough bread), plus Hmong-family farmers. Saturdays from 8 a.m. to noon, early May through late October (outdoors); and Sundays from 9 a.m. to 1 p.m., early November through mid-March (indoors).

Pennoyer Park Farmers Market

3601 7th Ave., (35th Street and 7th Avenue), Kenosha; 262-605-6700

If you love hopping on a bicycle to get your fresh veggies, then this market is your utopia because the market is located along Kenosha's Pike Bike Trail. It is also next to the city's bandshell, home to Kenosha Pops Concert Band's outdoor performances later that evening; and near Pennoyer Park beach (impromptu picnic?). Tuesdays from 6 a.m. to noon, June through November.

Pewaukee Farmers Market

Christ Lutheran Church parking lot, W240 N3103 Pewaukee Rd., Pewaukee; 262-737-6170

The market—sporting an artsy logo with a red rooster—features live music as well as vendors who are farmers and artists. You can find prepared foods like salsas and sauces (to pair with vegetables bought here, perhaps?) plus natural skin-care products. Wednesday from 3:30 to 7 p.m., mid-June through late September.

Port Washington Farmers Market

Main Street between Franklin and Wisconsin Streets, Port Washington; visitportwashington.com

Just a block from Lake Michigan's shoreline, in downtown Port Washington, vendors at this farmers market sell everything from mushrooms (Gourmet's Delight Mushrooms) and garlic (Grappling Gardens) to spices (A Spice Above the Rest) and honey (Little Mountain Apiaries), even jewelry from Jean's Rocks and Gems. Pick up eggs and pasture-raised meats from Burkel Family Farm, a fifth-generation family farm. Saturdays from 9 a.m. to 1 p.m., late June to late October

Racine Farmers Market—HMS

1012 Main St., Racine; 262-633-5796

Located across the street from Gateway Technical College, this market has been going strong since the early 1900s and now in a new location. You will find a mix of crafts and arts and fruits and vegetables (including from Gall's Garden Produce Farm in nearby Mount Pleasant), plus canned products, soaps and lotions, and plant starters—all made by artisans. Wednesdays from 11 a.m. to 3 p.m. and Saturdays from 8 a.m. to noon, early June to late October.

Riverwest Gardener's Market

North Pierce and Center Streets, Milwaukee; riverwestmarket.com

One of Milwaukee's most eclectic and diverse neighborhoods (Riverwest) expresses itself in true form with this market, where urban homesteaders, farmers, artists, and gardeners sell their wares. Bonus: this market is perfect for late risers. You will find gladiolas, sunflowers, and other cut flowers; La Tarte's fruit pies; jars of dill pickles and strawberry jam; bundles of dried herbs; produce like sungold tomatoes and red potatoes when in season, and more. Sundays from 10 a.m. to 3 p.m., early June through late October.

Saukville Farmers Market

249 E. Dekora St. (adjacent to Veterans Park), Saukville;
262-284-9423

The farmers market in this Ozaukee County town—which, back in the day, was surrounded by dairy farmers; then a stagecoach stop between Milwaukee and Green Bay—features many farmers who grow nearby, whether it is flowers or vegetables. Artists and crafters also sell their wares here. Sundays from 9 a.m. to 1 p.m., mid-June through late October (excluding Labor Day weekend).

Scio Farmers Market

Fairfield Plaza, 2133 Eastern Ave., Plymouth (outdoors); Fountain Park, 8th Street and Erie Avenue, Sheboygan (outdoors); visitsheboygan.com

Locals love to get cut flowers at these two markets, which started in 1989 and are organized by Sheboygan County Interfaith Organization, with an eye on educating about healthy-eating choices and supporting local farmers. Plymouth: Thursdays from noon to 5:30 p.m.; and Sheboygan: Wednesdays and Saturdays from 9 a.m. to 1 p.m., early June through late October. A winter market is in Sheboygan the first and third Saturdays from 8 a.m. to 1 p.m., November through May.

Sheboygan Farmers Market at Festival Foods

Festival Foods parking lot, 595 S. Taylor Dr., Sheboygan;
920-694-6260; festfoods.com

Unique to this market is that you can pop into the Festival Foods store (a major supermarket chain) before or after to pick up ingredients you will need to cook or bake with what you buy. Everything sold at the farmers market is grown by a local farmer and you do not have to worry about finding a parking spot (the lot is huge!). Ten Festival Foods stores around Wisconsin host farmers markets, in fact, including Appleton and Fond du Lac. Sundays from 7 a.m. to 1 p.m., mid-July through early October.

Shorewood Farmers Market

Estabrook Park, 4100 Estabrook Pkwy., Shorewood;
shorewoodfarmersmarket.com

This market's location is the perfect spot to bring a blanket or chairs for an impromptu picnic. Donut Monster's breakfast sandwiches and donuts often yield the longest line. Put in your order then shop Tree-Ripe's peaches and blueberries (when in season), Alice's Garden's herbs and herbal products, Chalkboard Kitchen's granola and, of course, vegetables and fruits from local growers, plus grass-fed beef, and Silver Moon Springs's salmon. You can even get your knives sharpened. Sundays from 9:30 a.m. to 1 p.m., mid-June through late October.

Southshore Open-Air Markets

Fountain Banquet Hall parking lot, 8505 Durand Ave., Sturtevant
Literally everything you need for that night's dinner can be picked up at this Racine County market, whether it is jam or preserves, sweet corn, tomatoes, beets, and squash. Yearn to own jewelry that no one else has? Antler Jewelry's necklaces are born out of elk and deer antlers. Bath bombs, candles, and CBD oil are also sold at the market. Mondays from 10 a.m. to 1 p.m. (June) and 9 a.m. to 1 p.m. (July through November).

South Milwaukee Downtown Market

1101 Milwaukee Ave., South Milwaukee; 414-499-1568;
smmarket.org
Live music—different acts very week, spanning bluegrass, folk, blues, and acoustic rock (5 to 7 p.m.)—arts and crafts, and Chinese food accompany shopping at this downtown South Milwaukee market. Items sold through fifty vendors include soy candles, farm-fresh eggs, apples (Apple Holler and Quednow's Heirloom Apples), pickles and peppers (Dave's Famous Pickles), olive oil (Marva's Greek Oil), and sausage (Milwaukee Sausage Company). Thursdays from 3 to 7 p.m., late July through early October.

South Shore Farmers Market

South Shore Park, 2900 S. Shore Dr., Milwaukee;
Southshorefarmersmarket.com

Shopping at this Bay View market is paired with view of Lake Michigan's ribbon of blue, along with live music and a peek into the past via a street organist. Get in line for your National Cafe breakfast sandwich first, then troll the stands for Door County cherries (when in season), Clock Shadow Creamery cheese curds, Hmong farmers vegetables and River Valley Ranch's exotic mushrooms. Prepared foods include whimsically flavored popsicles from Pete's Pops, Lopez Bakery tamales and Liege-style waffles from Press Waffles. Saturdays from 8 a.m. to noon, mid-June through late October.

Thiensville Village Market

Village Park, 250 Elm St., Thiensville; 262-242-3720

Spanning much of the day, which means you are not locked into a tight time frame, this market features vendors who offer everything from hot peppers to sunflowers, with fun finds like homemade dog treats. Bummed you are missing the gym to shop the market? No worries—a pop-up boot camp (from Burn Boot Camp) kicks off the market at 6:30 a.m. each week, on site. Tuesdays from 9 a.m. to 3 p.m., late June through mid-October.

Tosa Farmers Market

Hart Mills parking lot, 7720 Harwood Ave., Wauwatosa;
Tosafarmersmarket.com

Smack-dab in the middle of downtown Wauwatosa, the market's energy is lively with musicians, outdoor yoga, and cooking demos to get you jazzed about the weekend—and the fruits and vegetables (including from Kaleidoscope Gardens, an urban farm only four miles away), plants, prepared foods (from nearby Pete's Pops to Tabal Chocolates), meat, fish, dairy (such as Hoard's Dairyman Farm's cheese, from Fort Atkinson), and baked goods you are taking home. What sets this market apart: kids (ages five through twelve) are awarded $2 tokens to spend, learning about farm-fresh produce in the process. A Makers Market (selling pottery, textiles, soaps, jewelry, candles, and art) joins the market the first Saturday of the month. Saturdays from 8 a.m. to noon, June through mid-October.

Urban Ecology Center's Local Farmer Open House
First Saturday of March; Urban Ecology Center at
Riverside Park, 1500 E. Park Pl., Milwaukee; 414-964-8505;
urbanecologycenter.org

To help keep farmers in business and ensure their income's locked in before planting seeds, many offer community-supported agriculture (also known as CSA) subscriptions. Each week—or every other week, depending on how you customize—a box of what is harvested on the farm comes to you, usually via a local pick-up spot. Most subscribers are in cities and the farm, of course, is not, but events like chef dinners, harvest festivals, culinary workshops, and volunteer workdays allow members to see exactly where their food is grown.

But just like picking a doctor, hairstylist, or auto mechanic, choosing which farm to go with is a daunting decision. You want to meet the face behind the farm, right?

That is what motivated the Urban Ecology Center to launch the Local Farmer Open House in 2003, an annual event hosted each March. There is no cost to enter. And if you get jazzed about eco-friendly buildings, look around because this is among Milwaukee's best examples, with one of the largest solar power stations in the state; repurposed, rainwater flush toilets,and century-old hardwood maple flooring. As you wander from booth to booth, chatting up the farmers, as sunlight pours in through walls of windows, know that it is not uncommon for farms' CSAs to fill up at the end of this day. You will likely need to make a firm decision on your CSA that day. Think of it like speed dating: although brief, this interaction should allow you to make an emotional connection. And do not be fooled into thinking, *Oh, I'll just drop in, find a farm and leave.* You will easily be lured into what builds the schedule at this four-hour event. Cooking demos school in how to turn a CSA box into a week's worth of meals. Kids can participate in farm-themed games. Gardeners may want to learn about pollinators. Lunch can even be bought on site.

New to a CSA? Arrive just after the event begins for a workshop dubbed "CSA Basics," which lays the foundation for how a CSA model works—both for the members and the farmers—and provides tips on how to choose the best one for you and your family. There is another workshop offering in the afternoon for later arrivals.

Union Grove Public Market

Piggly Wiggly parking lot, 4400 67th Dr., Union Grove; 262-617-9922; uniongrovechamber.org

Despite its new name, this remains a market where farmers sell vegetables and fruits. The name change—and, as of 2020, new location—also reflects artists and crafters. Pick up dinner from a food truck, or a cool treat from Kona Ice, after you have scored items like honey, cut flowers, and candles, plus whatever produce is in season. A tarot-card reader and funnel-cake vendor are sometimes here, too. Tuesdays from 3 to 7 p.m., late June through late September.

Walker Square Farmers Market

Walker Square Park (between 9th and 10th Streets and Washington and Mineral streets), Milwaukee; walkersquare.org

This neighborhood tucked into Milwaukee's South Side is one of the city's most ethnically diverse—and this market's farm-fresh finds prove it. Sold by farmers are items like huitlacoche (corn mushroom, or Mexican truffle, a delicacy in Mexico), tomatillos (a Mexican tomato), and squash blossoms (serve sautéed) along with everything else you would expect at a farmers market, such as apples, sweet corn, potatoes, lettuce greens, and tomatoes. Thursdays from 7 a.m. to 2 p.m., early June through late October.

Watertown Farmers Market

Riverside Park, 850 Labaree St., Watertown; 920-342-3623; Watertownmainstreet.org

One of the state's longest running farmers markets—this one began in 1862 as Watertown Fair Day—sells a variety of foods crafted by artisans along with farm-fresh produce. This includes Good Morning Babies Beef's grass-fed beef, Oliver Twists' dog treats, Lucky Penny Coffee (in a cute 1970s Shasta camper), Prairie Pure Cheese's cheddar and Butterkäse, and produce from Crimson Kitchen & Gardens (a Watertown grower). Tuesdays from 7 a.m. to noon, early May to late October.

Waukesha Farmers Market

125 W. St. Paul Ave., Waukesha; 262-547-2354; waukeshafarmersmarket.com

With 150 vendors throughout the season, the market in downtown Waukesha features vendors you will not find at other area markets, such as Cupcakes by Jordan James, jars of pickles from Muddy Boots Plants & Produce, Herbal Herbalist's soy-wax rainbow-striped candles, and Mallory Meadows's hemp. Several nearby businesses also have a stand, like Allô! Chocolat. There is also live music. Saturdays from 8 a.m. to noon, early May through late October.

West Allis Farmers Market

6501 W. National Ave., West Allis; westallisfarmersmarket.com

One of the Milwaukee area's best farmers market—open three days each week, with a longer season and many vendors—sells products like antique apples, CBD bath and body products from a female-owned ten-acre hemp farm, starter plants, dairy (cheese and milk), and Hmong families' farm-fresh produce. Grab bubble tea or a cup of coffee from vendors along Lapham. Tuesdays and Thursdays from noon to 6 p.m.; and Saturdays from 1 to 6 p.m.; early May through late November.

West Bend Farmers Market

Old Settlers Park, 200 N. Main St., West Bend; 262-338-3909; downtownwestbend.com/farmers-market.html

Around eighty vendors come together to sell kitchen staples—fruits, vegetables, meats (lamb, beef, poultry, turkey, and elk), and dairy items like eggs and cheese—at this market but it also attracts those selling items you may not expect, such as ice cream, fish, spices,

teas, and coffee. Everything you may need to make an amazing Saturday-night dinner can be bought in one place. Hanging floral baskets for sale add further ambiance. Saturdays from 7:30 a.m. to 11 a.m., late May through early October.

Westown Farmers Market
Zeidler Union Square, 301 W. Michigan St., Milwaukee; 414-276-6696; westown.org

Downtown office workers love to drop by this market—sponsored by the Westown Association, on the city's oldest public park—on their lunch breaks (the fifteen prepared-food options are stellar, like Falafel Guys and Little Havana Express). Alice's Garden Urban Farm and Teens Grow Greens—two of the fifty to sixty vendors each week—teach inner-city residents to grow their own food. Live music spins out from the park's gazebo. Wednesdays from 10 a.m. to 3 p.m., June through October.

Whitefish Bay Farmers Market

325 E. Silver Spring Dr., Whitefish Bay

From the Mushroom Lady's foraged *fungi* (like Chicken of the Woods) to Chillwaukee's inventive flavored popsicles, this petite-sized farmers market pops up in the Aurora parking lot along bustling East Silver Spring Drive. Staples like farm-fresh eggs, just-cut flowers, and fruits and vegetables (including antique apple varieties from Quednow's Heirloom Apple Orchard; and Larryville Gardens's "naturally gardened" herbs, heirloom tomatoes, eggplants, and more started from seeds directly from Italy, designed to help shoppers cook rustic Italian cuisine) round out the vendors' offerings. Saturdays from 8 a.m. to noon between mid-July and mid-October.

Whitewater City Market

Historic Train Depot, 301 W. Whitewater St., Whitewater (outdoors); and Irvin L. Young Memorial Library, 431 W. Center St., Whitewater (indoors); 262-473-2200; downtownwhitewater.com

Since this year-round market debuted in 2015, it is attracted around sixty vendors—plus local craft beer, lawn games, food carts,and music—to the grounds of a Gothic-style historic train depot (home to the local history museum) during the warmer months and, come winter, about twenty vendors at an indoor location. Tuesdays from 4 to 7 p.m., May through October. Saturdays from 9 a.m. to noon, November through April.

Whitewater Farmers Market

Depot Museum parking lot, 301 W. Whitewater St., Whitewater; 262-949-4508; whitewaterfarmersmarket.com

This market debuted in 1991 and offers everything from Jeanie's Sweets' vegan muffins and apple-cider donut holes to Bluff Creek Nursery's perennial flower plants, plus honey, vegetables, grass-fed beef, melons, and Murphy's Sweet Corn's raspberries and sweet corn. Strawberries and asparagus are examples of what is available earlier in the season, in June. Saturdays 8 a.m. to noon, early May through early November.

Apple Barn Orchard & Winery

W6384 Sugar Creek Rd., Elkhorn; 262-728-3266;
applebarnorchardandwinery.com

Now on its seventh generation, this orchard can be traced back to
Jacob and Maren Brensetre, who emigrated from Oslo, Norway, in
1846, smoking hogs in a smokehouse still on the property. During
the 1980s the family planted apple trees—there are now four
thousand—and later launched a fruit winery in 2004. Pick-your-own
apple varieties include Honeycrisp, Cortland, McIntosh, and Jona
Gold. Apple treats and apple cider are sold in the bakery. Strawberries and pumpkins can be picked when in season.

Apple Holler

5006 S. Sylvania Ave., Sturtevant; 262-884-7100; appleholler.com

Driving along I-94 between Milwaukee and Chicago you literally cannot miss this seventy-eight-acre attraction. Particularly in autumn,
dozens of cars are parked along the frontage road and if you lean
out the car window, you may catch a whiff of baked apples.

This orchard, owned by the Flannery family since 1987, grows
and sells apples (including beloved Honeycrisp, plus Zestar and Ginger Gold), peaches (Sentry, Red Haven and Desiree are among the
varieties), and pumpkins. It is akin to a mini–Cracker Barrel. Looking
for items to stock your fridge and pantry, or some décor that shows
off your love for farms? This is that place, with a robust gift shop and
bakery. Yummy treats sold in the bakery include apple-cider donuts;
apple pies; apple turnovers; and apple butters, jams, and jellies.
Additional goods like salad dressing, pretzel dips and salsas—plus
Wisconsin wines—can also be bought here.

What makes Apple Holler different from other farm stands
is it is open year-round. This includes the sit-down cafe, called
Applewood Grill, open for breakfast, lunch, and dinner. It is where
fluffy apple-buttermilk pancakes topped with slices of warm,
cinnamon-laced apples are served. They also offer a daily fish fry.

At the Orchard Market, you can get more than a slice of apple pie: cherry-cider donuts, caramel apples and apple fritters are just three additional options. Tack on flavored apple cider (amped up with cinnamon, cranberry, caramel, or gingerbread) to your order, too.

Picking packages are offered when apples and peaches are in season. The Flannery's practice integrated pest management so as not to harm the land. Keep an eye on Apple Holler's website for updates to its picking calendar. Pumpkins can be selected straight out of the pumpkin patch come fall when activities celebrating harvest—with a farm twist—include hayrides, and gemstone and fossil mining (geared for kids). During June, July, and August, a daily admission fee (which includes an Orchard Express train ride) allows for petting farm animals, kids' play areas, a Jack and the Beanstalk Story Trail, and a cow maze. This farm is even open during the winter, with sleigh rides and visits with Santa Claus and his reindeer.

Basse's Taste of Country

3190 County Line Q, Colgate; 262-628-2626; bassesfarms.com

If it is mid-June at this farm (in business since 1997), it is the start of strawberry-picking season—celebrated at the strawberry fest, where strawberry donuts are served—followed by raspberries and blackberries later in the summer. Come fall a corn maze—pardon the pun—crops up along with the sweet-corn harvest, an apple and sunflowers festival (includes a three-acre sunflower maze), then a pumpkin fest. A calendar on the farm's website outlines picking dates along with hours. There is even a "U-pick strawberry" hotline.

Barthel Fruit Farm

12246 N. Farmdale Rd., Mequon; 262-242-2737; barthelfruitfarm.com

In 2019, Sue and Jeff Knudsen bought this orchard, in the Barthel family since 1839. She returned home after culinary school in Connecticut while he is from Vermont. Pick apples, pumpkins, and strawberries. The Knudsens also harvest pears, gourds, and sugar-snap peas. Consider clearing out your car: a vast variety of annual flowers, herbs and vegetable plants are sold; and a cute bakery truck (all Sue's doing) showcases what you can make with farm-fresh fruits, such as apple-cider donuts and mini-pumpkin cheesecakes, plus apple-cider and strawberry-lemonade slush.

Brehmer's U-Pick Strawberries

5805 Clover Rd., Hartford; 262-673-6527

Although you can score the Brehmer's strawberries at the Fondy Food Market, it is much more fun to pick those berries yourself, right? The one-and-a-half-acre strawberry patch is open for picking mid-June through early July. This family-owned farm also grows a bunch of other vegetables, including tomatoes, corn, hot and sweet bell peppers, kohlrabi, and red and white potatoes, all sold at the farm through late fall. Stay tuned to the growing calendar—including picking in the pumpkin patch—via the farm's Facebook page.

The Fideler Farm

2863 Ridge Rd., Kewaskum; 262-338-0494; thefidelerfarm.com

Nestled in the Kettle Moraine State Forest's Northern Unit, this seventy-acre family-owned farm (in the same family since 1979) invites you out to pick strawberries in early summer, and also operates a farm stand. You can find their berries, jams and jellies, dilly beans, farm-fresh vegetables and fruits they grow, and delicious delights like raspberry-cheesecake pies at the Tosa, Kewaskum, and West Bend farmers markets.

Henke's U-Pick Raspberries

1987 W. Mill Rd., Jackson; 272-677-4672

This family-owned farm in Washington County opens for raspberry picking each year beginning in late August or early September (depending on the season) and extending through late October (or the first frost, whichever comes first). Keep an eye on its Facebook page for news of that season's ripening dates so you do not miss out.

Mayberry Farms

W2364 County Road Y, Mayville; 920-387-3696; mayberryfarmswi.com

Tim and Danielle Clark bought this sixty-acre farm in 2017, shortly after they married, and are raising three young boys. Strawberries represent the bulk of their sustainably farmed crops (they also produce raw honey and grow flowers and open up the sunflower fields for photo shoots with photographer partners) and there is an option to pick your own strawberries, shop at its store, and hang out at picnic tables for lunch before or after.

Peck & Bushel Fruit Company

5454 County Road Q, Colgate; 414-418-0336; peckandbushel.com

This family-owned, sseventy-acre apple orchard (its thirty thousand organically farmed trees yield Honeycrisp, Cortland, Sweet Sixteen and Macoun, among others) in the town of Erin is so charming that couples rent the three-thousand-square-foot, white and modern

barn for their wedding-day celebrations. Twenty-five different apple varieties are grown here and picking (weekends only) begins in late August. The farm, which celebrated its 10th-anniversary in 2020, also presses apples into cider and sells pre-picked apples in the barn. Tours are self-guided.

Polzin Farms
1758 Hwy. I, Grafton; 262-375-3276

Offering both pre-picked and you-pick options from its strawberry fields each summer, this farm also grows and sells sweet corn, beans (both green and yellow), carrots, lettuce, dill, pickles, zucchini, cucumbers, and onions. In the summer of 2020 the family built a stand to their farm, which is in the town of Cedarburg, and is also known for its unique popcorn flavors. Meet the Polzins at their stand at Milaeger's Great Lakes Farmers Market, hosted each Sunday at Milaeager's Racine store, and farmers markets in Brookfield, Thiensville, and West Allis.

Thompson Strawberry Farm
14000 75th St., Bristol; 262-857-2353;
thompsonstrawberryfarm.com

I am in the strawberry fields with my mother, plucking berries off bushes that are as tall as my scabbed, sunburned knees. It is a hot afternoon in July during the late 1970s and my brother will be born any day now. My mother's wearing a red gingham-check shirt and her belly is swelling with the news that, soon, I will no longer be an only child. My brother is about to come roaring into this world, four years after me, but for now we are filling the time with one of the Midwest's most treasured summer activities: picking fruit at a farm.

I am not the only kid who has fond memories of picking fruit at this family-owned fruit farm, just over the Illinois state line in Kenosha County. It is a forty-five-minute drive from Milwaukee, an hour north of Chicago. The farm dates to 1969 and celebrated its fiftieth anniversary in 2019.

Is that afternoon, so firmly imprinted in my memory even decades later, the reason I felt drawn to supporting boutique farm businesses as an adult? Did gently dropping the just-plucked berries into a box or bucket (my child's mind cannot remember) teach me how crucial it is to know where your food comes from, to actually think about the process—dare I say, the *journey*—in which it arrived in your hands?

Years later, on a cruise ship in the Caribbean, on a tour of the bridge where the command center keeps the ship right, I spotted a banana boat trudging along from Colombia to South Florida. I can no longer eat a banana and not have that vision wash over me, the thought of food traveling thousands of miles, across an ocean, to reach my breakfast table.

May we all have the joy of picking fruit and bringing it home, suddenly faced with the delightful problem of "What now?" and turning on the oven to bake muffins, plopping sliced fruit into pancakes, ladling onto scoops of vanilla ice cream.

Wunberg Produce

W9219 Lakeshore Rd., Sharon; 715-415-0585; wunbergproduce.com

Merging their last names, Rob Wesenberg and Kenneth Wundrow came up with Wunberg for their twenty-five-acre fruit farm (growing strawberries, raspberries, blueberries, blackberries, elderberries, and cherries; plus, apples, pears and pumpkins), a business that includes an acre on a family member's farm. Rob grew up milking cows on a farm while Kenneth works as a tax advisor (but always wanted to farm). They also produce and sell, under their own label, honey, brown eggs, maple syrup, dill pickles, and jams. You-pick strawberries are available at their farm each June.

Tree Farms

Buffalo Bill's Christmas Trees

9612 W. Oakwood Rd., Franklin; 414-427-5155; buffalobillschristmastrees.com

At this rare tree farm in Milwaukee County, the owners create fun events like Bring Your Pet Day (dogs are always welcome and treats available, but this is your chance to dress up Fido in holiday garb). Each year between 1,500 and 2,000 trees—Canaan Fir, Douglas Fir, Blue Spruce, White Spruce, and Scotch Pine—are planted for a grand total of around 11,000. Free hot chocolate and cider, pine-cone crafts for kids, and sledding (if there is snow) turn a tree-cutting trip into a festive outing.

Poplar Creek Tree Farm

3895 S. Woelfel Rd., New Berlin; 262-442-1275

Looking for a backdrop for your holiday-card photo? Contact the farm for details (even leashed dogs can come). In fact, the farm's Facebook page proves dogs are welcome, with nearly as many photos of trees as visiting dogs. Randy and Karen Cooper transformed Karen's family's land into a Christmas-tree farm in 1994 (they are the fourth-generation owners), planting Scotch Pine, Balsam Fir, and Canaan Fir. This pick-your-own tree farm also sells wreaths and Christmas-tree stands.

Sandhill Tree Farm

2323 E. River Rd., Grafton; 262-573-7206

Rick and Lisa O'Malley met on a blind date. They both owned farms—Trees for Less Nursery (Rick) and the future site of their Christmas-tree farm (in Lisa's family since 1963). "The [sandy, loamy] soil [along the Milwaukee River] is absolutely perfect for growing Fraser Fir, very unusual in Southeastern Wisconsin," says Lisa, "and every year we provide the Governor's Mansion with four trees." After cutting a tree, visitors can say "cheers" with hot chocolate or a

© Sandhill Tree Farm

Jameson shot by the firepit . . . or under the mistletoe. At press time, ownership was transferred to the Ness family.

Sugar Creek Tree Farm
N6447 Church Rd., Burlington; 262-767-1177
The Cheskys' thirty-two acres spawn six types of Christmas trees: Fraser Fir, Scotch Pine, White Pine, Colorado Blue Spruce, White Spruce, and Balsam Fir. All are available for cutting. Do not be shy about arriving hungry or thirsty as you can easily whet your palate with hot chocolate or apple cider along with cookies (food and drink on weekends only).

Trees For Less Nursery
11550 N. Wasaukee Rd., Mequon; 262-242-7522; treesforless.com
While the land has been in the O'Malley family since 1948 (when owner Richard O'Malley's grandfather bought it), the nursery expansion debuted in the late '70s, followed by the planting of twenty acres of Christmas trees in the early 2000s. Cut your own Balsam Fir, or pine and spruce varieties, or pick out a pre-cut Fraser Fir. Maple syrup crafted on the farm is also sold, and hayrides, hot chocolate, snacks, and a large fire pit make the trip even more festive.

Wagner's Tree Farm
W1516 WI-60 Trunk, Rubicon; 262-227-0601; wagnerstreefarm.com
Ed and Pam Wagner invite you out to cut a Christmas tree (choose from six varieties, including Norway Spruce, Black Hill Spruce, and Balsam) at their farm, in the family since 1960. Pre-cut options (expanded to include White Pine and Canaan) are also available. Picture nights with Santa are hosted throughout early December. Drop into Rosemary's Gift Shop, tucked into a barn, for hot chocolate as well as items to buy like soy candles and holiday décor.

Armstrong Apples, Orchard & Winery

W853 County Road B, Campbellsport; 920-477-3007;
armstrongapples.com

Yoga classes (followed by a glass of wine or cider) and comedy-club evenings turn the idea of a farmstead on its head. Taste their fruit and Wisconsin-grown grape wines (or, on a hot day, frozen wine slushie or frozen apple cider) and hard apple cider, then order a Bavarian-style pretzel (with spicy apple sauce) right out of the oven. Apple products for sale include pies, caramel apples, jams, and cider. Come fall, apple sling shots keep kids entertained.

Aeppeltreow Winery & Distillery

1072 288th Ave., Burlington; 262-496-7508; aeppeltreow.com

At Brightonwoods Orchard, where two hundred antique and heirloom apple varieties are grown, walk into the barn (this is Aeppeltreow: in Old English this means "apple true") and taste small-batch cider, perry (pear-based cider), and spirits, crafted from apples and pears, along with whiskey and apple brandy. Charles McGonegal launched the winery and distillery, and the tasting room is open weekends only in May, June, November, and December; Friday through Sunday in July and August; and daily except for Monday in September and October.

Duesterbeck's Brewing Company

N5543 County Road O, Elkhorn; 262-729-9771; dbcbrewery.com

Giving her family's farm (dating back to the late 1800s) a new lease on life, Laura Duesterbeck Johnson and her husband Ben Johnson launched their microbrewery in late 2019. They even built a new barn to replicate the original one, but with modern conveniences. This is the kind of barn you never want to leave, with pizzas, giant soft pretzels, cheese curds, summer sausage and other snacks served in the taproom, plus kombucha and Sprecher sodas for the designated driver. Around fourteen different Duesterbeck' beers are on tap at once.

Edwin Brix Vineyard & Sell Family Wines

N4595 Welsh Rd., Juneau; 920-219-4249; edwinbrix.com

Except for its E Flat Eddie, folding in Kirschbaum's strawberries, every wine is crafted from grapes grown on the Sells' farm, now on its fifth generation and having farmed different crops over time. Although the first wines were made with New York–grown Marquette grapes, the family pivoted to their own grapes in 2013. Their first vintage (2014) was followed by the tasting room's 2019 opening, providing an opportunity to sip through this family's most recent chapter, which includes 11 wines crafted from cold-hardy grapes like Sabrevois and Brianna.

Pieper Porch Winery & Vineyard

S67 W28435 River Rd., Mukwonago; 262-349-9092; pieperporchwines.com

The visitor experience at Pieper Porch, located only nine miles from downtown Waukesha, ranges from what you would expect (tasting through its forty-some wines) to fun twists like wine slushes dubbed Black Currant and Phantom Princess. Locals unwind here with friends on the weekends (the only time the tasting room's open), thanks to its namesake porch with a vineyard view. The owners encourage bringing your own food. Wines are made with a mix of traditional wine grapes (such as Malbec, Pinot Noir, and Chardonnay) and fruit (including Door County cherries and Warren cranberries).

Solu Estate Winery & Meadery

W8269 County Road F, Cascade; 920-528-1550; soluestate.com

With a nightlife scene—come summer, its live-music series at a pavilion features locally loved acts, like The Whiskeybelles—SoLu Estate Winery is an hour's drive north of Milwaukee and thirty-five miles from Fond du Lac, nestled in the Kettle Moraine National Forest. Music concerts are Thursday, Friday, and Saturday evenings, plus Sunday afternoons, on the outdoor patio. The winery opened in 2016 and its name is a combination of the owners' kids' names: Sophia and Lukas.

On the weekends, the winery's tasting room and twenty-six-acre vineyard host events that attract city folks keen on a short drive to a pastoral setting. This includes serving up yummy, inventive flatbreads topped with ingredients like cherries, BBQ pork, and mushrooms, and it is not unusual to have a food truck (such as That Taco Guy, a veteran-owned business) park here for additional eats. Owners TJ and Anne Sommer continually work hard to invite locals to share their creative talent. Recently that meant a local photographer capturing shots of visitors (plus their dogs), in exchange for tips. There are also weekday events. Each Wednesday's farmers and crafters market curates a mix of vendors who sell their food and art.

Grown in the vineyard—which was planted in 2003 and first used to produce wines in 2016, coinciding with the tasting room's opening—are cold-hardy grape varietals like St. Croix, Marquette, Frontenac Gris, Brianna, and St. Pepin. Grapes grown from outside of Wisconsin, in regions celebrated for viticulture, are often used, too. Mead wines are also made by the winery. Fun, easy-to-remember names like Amazon Blue (a mead folding in blueberries and açaí berries), Wild Ginger Root (another mead with ginger accents the owners say drinks like a carbonated ginger ale), and Venus Dry Rosé (estate Frontenac Gris grapes) are three examples. Inventory shifts based on season and availability.

Spirits of Norway Vineyard

22200 W. Six Mile Rd., Franksville; 414-430-0128; spiritsofnorwayvineyard.com

Although this winery (in the town of Norway) does not operate a tasting room, visitors are welcome to the vineyard—which grows cold-hardy grapes for its Estate Collection like Brianna, Marquette, Frontenac Blanc, Somerset, Petite Pearl, Marechal Blanc, and Catawba—on the third Saturday of July. Grapes that do best in other climates are sourced, too, such as Chardonnay, Merlot, Syrah, Trebbiano, Montepulciano, and Barolo. Farm-to-table events are hosted as well. If you live nearby, delivery is free within twenty miles with a minimum four-bottle purchase.

Staller Estate Winery & Vineyard

W8896 County Road A, Delavan; 608-883-2100; stallerestate.com

Partnering with a local chef on five-course dinners and offering picnic platters, charcuterie boards, and a "wine and chocolates" flight tasting . . . is this Sonoma? That is precisely the vibe—all about lingering—owners Wendy and Joe Staller strive to create, just without the California grapes. Their property, not that far north of the Illinois-Wisconsin state line (about twenty miles), is just twenty minutes from Lake Geneva and fifty-five miles southwest of downtown Milwaukee.

"We use all cold-climate, winter hardy varietals that can be grown in the Midwest," says Wendy, such as La Crescent, Maréchal Foch, and Frontenac, grown in their three-acre vineyard, and viewable from the winery, particularly if seated outdoors. Resembling a red barn, it fits right into this agricultural region.

The Stallers—who met as science students at the University of Wisconsin-Whitewater and both grew up nearby, with Joe's roots dating back to his family's dairy farm—debuted their winery in 2007 with a seven-thousand-bottle vintage made from contracted grapes. Wendy serves as the winemaker, having schooled herself on oenology through the leading wine program in the country, at the University of California-Davis. Staller Estate's wines have racked up awards at competitions like the Finger Lakes International Wine & Spirits Competition (in New York). Names are a mix of quirky (Lady in Red, a semi-dry red table wine; or Whitewater Rush, a semi-dry white) and those that celebrate the cold-hardy varietals, like the Blanc De Crescent and Reserve La Crescent (both are crafted from La Crescent grapes).

During the summer, live music is hosted at the winery on Saturdays. The winery is also rented out for bridal showers, bachelorette parties, weddings, and corporate retreats.

Tons of wine-themed items to use at home or in the kitchen are sold in the tasting room, in addition to their wines, and tours are offered for those who wish to go deeper. This tends to interest home winemakers but also pacify cynics who simply can't believe Wisconsin wine not only can be made here but is also decent. The Stallers have brought home numerous awards for their wine, proving those folks wrong.

SOUTHWEST WISCONSIN

The bulk of Wisconsin's boutique farms are tucked into the Driftless Region's Vernon County, which attracts young couples on a new venture while keeping family farms that stretch back several generations still in business. (Fun fact: At around two hundred farms, the country's highest concentration of organic farms is in Vernon County.) During the Ice Age, glaciers "skipped" over this area across southwestern Wisconsin, southeastern Minnesota, and parts of Iowa, too, avoiding a flat fate. Tucked into the rolling hills and river valleys is a healthy mix of innovation in this farming community, like the Engel brothers' cold-pressed sunflower oil, rubbing shoulders with tradition (such as raising heritage hogs) and the quirky businesses keep travelers coming. Wisconsin Foodie host and James Beard Award–nominated chef Luke Zahm's Driftless Cafe is a veritable love letter to all the farmers nearby.

Farmers not just here but also in Green County host workshops in everything from floral arranging to beekeeping and if you want to learn more, a handful of folk schools coach in nearly lost arts such as blacksmithing and basketry (Driftless Folk School, Folklore Village, and Shake Rag Alley Center for the Arts).

What I affectionately dub "The Swiss Belt" is that sweet spot in Green County where cheesemakers rack up international awards by merely continuing in the path of their emigrant elders, who learned to make cheese in Switzerland. I have not yet met a Wisconsinite who does not take it for granted that alpine-style cheese is sold at local farmers markets alongside honey, meats, vegetables, fruits, and more. And if you should find yourself there, say hi to John and Lisa at Inn Serendipity in Browntown, one of the country's first eco-friendly farm stays that strives to teach guests all about homesteading.

Speaking of farmers markets, the proof of this region's robust farming culture lies at Dane County Farmers Market every Saturday morning. This is the country's largest producer-only farmers market. Shoppers arrive to the lawn surrounding the State Capitol Building as early as 6:15 a.m. to score scones baked by the Amish or Willi Lehner's cave-aged, bandaged cheddar. Foot traffic is so fierce that you must walk in one direction only around the market—and shop quickly, before it is all scooped up for the day.

Less than a mile away? The University of Wisconsin-Madison campus, where many of these farmers once attended classes or received encouragement in their farming careers.

Cheese

Green County Cheese Days

cheesedays.com; and greencountybarnquilts.com

Bruce Workman of Edelweiss Creamery in Monticello told me something I will never forget, which is that Green County's fair is like an international cheese competition. There is a reason: Some of the world's best cheesemakers, with their Swiss roots, live and farm here. The cheese they produce, which earns awards at international competitions, is mostly alpine style.

Each September, for an entire weekend, Green County Cheese Days celebrates this slice of the Old World. You may feel as if you have been airdropped into Switzerland. Locals flaunt their pride for the region—and their heritage—in the Swiss Colony Cheese Days parade, sponsored by Swiss Colony, a Monroe retailer of Wisconsin cheeses, sausages, and snacks that ships worldwide, and has been family-owned since 1926. The festival's mascot, dubbed "Wedgie," prances around and has his own song and dance, to the tune of "Roll Out the Barrel." Each year a king and queen, and prince and princess, are crowned, county-fair style, only they are judged based on their passion for the festival—not their youth and beauty. And, of course, there is a cheese tent showcasing wedges from producers like Landmark Creamery (sheep's-milk cheeses), Edelweiss Creamery (Emmentaler Swiss and aged Gouda), and Emmi Roth USA (founded in 1990 by a family-owned creamery in Switzerland dating back to 1863, its Grand Cru® Surchoix was named World Champion at the 2016 World Championship Cheese Contest).

A highlight of the weekend, particularly for those who want to learn more about farming, is a tour to a local family-operated dairy farm. The ride also takes you past the famed Green County Barn Quilts, where each barn's façade is emblazoned with a signature quilt-like design, chosen by the owners to reflect their heritage. For example, Bruce and Judy Meier's "Star in the Valley" quilt design on their Monroe barn is an homage to Bruce's uncles who initially farmed it, painted by Bruce's daughter's as a Father's Day gift in 2010.

Like any festival, there is no shortage of live music, food stands, and carnival games—only you may glimpse Old World influences like alphorn entertainment (playing a long horn as the Swiss do) and yodeling.

Another reason to head to Green County? The country's only remaining Limburger producer is in Monroe, at Chalet Cheese Cooperative. Taste this stinky, but delicious, cheese for yourself (in a sandwich) at Baumgartner's, snug on the town of Monroe's square.

Cider Tastings

Brix Cider
119 S. 2nd St., Mount Horeb; 608-437-2749; brixcider.com
Calling itself a "start-up cidery," which reflects the entrepreneurism of Marie and Matt Raboin, this cidery's mission—to source apples from local farms while still using the one hundred or so varieties grown on their eighty-acre Barneveld orchard (with a thousand trees)—helps keep the state's fruit-farming community sustainable. A map on its website depicts exactly where the apples, as well as food products for its culinary menu (such as Seven Seeds Farm's beef, Tietz Family Farms' popcorn, or Sartori's cheese), are sourced. Taste the goodness for yourself at its Mount Horeb tasting room.

Crawford County Fair

Crawford County Fairgrounds, 17725 WI-131, Gays Mills; 608-412-4748; crawfordcountywisconsinfair.com

Events here range from a grilled-cheese sandwich contest (do not smirk: the rules about bread and cheese choices are strict!) to truck and tractor pulls (including one just for antique tractors), plus carnival rides, a cake revue, a demolition derby, animal exhibits and shows (cats and dogs join fuzzy alpacas, rabbits, swine, and more), and live music. A fair ambassador is named each year. When: late August

Dane County Fair

Alliant Energy Center, 1919 Alliant Energy Center Way, Madison; 608-291-2900; danecountyfair.com

In addition to animal viewing areas and exhibit halls, Dane County Fair—with a goal of providing agricultural education to urban attendees—hosts everything from magician Lady Houdini to a new event called Rockstars in the Ring, where youth with an intellectual disability show livestock in a show just for them. Carnival rides, a kiddie tractor pull, nightly live music and Wisconsin-specific food (such as Dane County Pork Producers' pork-loin sandwiches and Stoughton FFA alumni's cream puffs) are also on the schedule. When: mid-July

Dodge County Fair

Dodge County Fairgrounds, Wisconsin 33, Beaver Dam; 920-885-3586; dodgecountyfairgrounds.com

What is unique about this fair is that most events (tractor pulls, concert, and demolition derby) are free with paid gate admission. The grandstand attracts national touring rock and country acts, and the music schedule also includes local bands and three family-friendly stages. Yummy snacks like funnel cakes and deep-fried cheese curds, plus carnival rides, are fun for all ages. Five animal barns and arena shows are joined by the educational Moo Barn and Netwurx Junior Dairy Barn. When: mid-August

Grant County Fair

Grant County Fairgrounds, 916 E. Elm St. (County Road A), Lancaster; 608-723-2135; grantcountyfairwi.org

From truck and tractor pulls—plus the quirky ATV and UTV garden-tractor pulls—to sheep showing, there is something for everyone at this fair, organized by the University of Wisconsin Extension. Carnival rides and family-specific entertainment (including a butterfly tent and crab racing) keep little ones engaged. The fair also hosts Junior Fair Livestock auction and exhibit halls for community members' arts and crafts (from oil paintings to clothing). When: mid- to late August

Green County Fair

Green County Fairgrounds, 2600 Tenth St., Monroe; 608-325-9159; greencountyfair.net

Cheese contests and judging—and, most importantly, tasting—are what sets this fair apart, which has been ongoing since 1857. Some of the world's best cheesemakers are in Green County. Also judged are animals, crafts, quilts, plants, flowers, and farm-raised vegetables. Each year the Fairest of the Fair (fair ambassadors) are crowned, and there are carnival rides—such as the Great Gondola Wheel—too. When: late July

Iowa County Fair

Iowa County Fairgrounds, 900 Fair St., Mineral Point; 608-987-3490; iowacountyfair.org

For over 170 years, this fair has filled with nostalgia like a livestock auction, watermelon-eating contest, community spaghetti dinner, and story time and ice-cream treats with the Fairest of the Fair, with new additions over the years like hypnotist performers. There is also no shortage of viewing animals like rabbits, sheep, and swine in buildings dedicated to each breed. When: Labor Day weekend

Juneau County Fair

Veteran's Memorial Park, 1001 Division St., Mauston;
608-547-2426; juneaucountyfair.com

Since 1865, three years before Wisconsin became a state, this fair has been a staple for locals' summer plans. From judging of animals (dogs, cats, exotic animals, rabbits, swine, you name it) to live music and a beer garden, plus a rodeo, crowning of Fairest of the Fair, and offbeat events like "dress the horse" parade, there is something for everyone. Be sure to grab some eats at the 4-H food stand. When: early to mid-July

Lafayette County Fair

Lafayette County Fairgrounds, 701 E. Louisa St., Darlington;
608-776-4828; lafayettecountyfair.org

Food here rocks, such as locally produced deep-fried cheese curds, Knights of Columbus cheeseburgers and ice cream. Events at the grandstand include tractor pulls and a combine derby. Live music is also featured each year, in addition to carnival rides, crafts judging, and animal viewing (such as swine barns or rabbits in the small-animal building). It is a "small and mighty fair," say fair organizers, and the tractor pull has been voted one of Wisconsin's best. When: second week in July

Lodi Agricultural Fair

Lodi Fairgrounds, 700 Fair St., Lodi; 608-592-4499; lodiagfair
.com

The Lodi Agricultural Fair has operated independently for over 155 years. Old and new fill the schedule, such as beer-garden karaoke and a polka band, or an old-fashioned horse pull and Celebrity Pie auction. Animal viewing areas cover typical breeds, and arts and crafts are celebrated with the Quilt Turning, in which stories behind one hundred quilts on display are shared. Carnival rides, Farmer for a Day (for kids), a restored one-room schoolhouse, horse-drawn carriage rides, and yummy fair food are other draws. When: early September

Rock County 4-H Fair

Rock County Fairgrounds; 1301 Craig Ave., Janesville;
608-755-1470; rockcounty4hfair.com

What makes this fair special is that it is the country's oldest 4-H
fair (over ninety years) with lots of opportunities for youth to show
animals. Funnel cakes, popcorn, cotton candy, caramel apples, and
ice-cream cones are among the fair food's sweeter selections, plus a
Wisconsin staple: fish fry. When: early August

Richland County Fair

Richland County Fairgrounds, 23630 Co. Hwy. AA, Richland
Center; 608-647-6859; fair.co.richland.wi.us

This fair has evolved since its beginning and mixes newer attractions,
like a beer garden (also selling brats and burgers), with traditional
fair delights like carnival rides (including a gigantic Ferris wheel), a
demolition derby, truck and tractor pull, tiny-tractor pulls for kids,
crowning the Fairest of the Fair, and animal judging. When: early to
mid-September

Sauk County Fair

Sauk County Fairgrounds, 700 Washington Ave., Baraboo;
608-356-8707; saukcountyfair.com

This fair, organized by the Sauk County Agricultural Society, has
been going strong since 1855, continuing to showcase farm animals
(plus more domestic furry creatures, like cats and dogs) and pro-
mote further awareness of agricultural life. Whether you are viewing
the fair from atop a carnival ride or at the Grandstand (dating back
to 1936) for an antique tractor (or truck) pull or rock band's show,
many activities are packed into the five-day fair. When: mid-July

Trempealeau County Fair

Trempealeau County Fairgrounds, 19780 Park Dr., Galesville;
608-582-4508; trempealeaucountyfair.com

Organized by the Trempealeau County Agricultural Society, the fair's
foods include fried mini donuts, walking tacos, cheese curds, funnel
cake dusted with powdered sugar, and milkshakes. Local crafters
and artists exhibit their wares and, of course, youth and adults

demonstrate their husbandry skills via livestock shows. The tractor pull and demolition derby are always a huge draw. When: mid-July

Vernon County Fair
Vernon County Fairgrounds, Highways 14 and 61, Viroqua; 608-637-3165; vernoncountyfair.com

The Driftless Region's allure for many farmers (especially those who are operating boutique farms and farming sustainably) makes this county fair a huge draw. But the carnival rides are also "one of the best in the state," say fair officials, who call this "a simple county fair," but not in a way that is demeaning. There is lots of nostalgia at this fair, including the 4-H-operated food stand, draft horse show, and multiple animal-viewing areas. When: mid-September

Dairy Centers

Babcock Dairy Plant and Dairy Store
1605 Linden Dr., Madison (Babcock Hall); and 800 Langdon St. (Memorial Union), Madison; 608-262-3045; babcockhalldairystore.wisc.edu

It is impossible for me to *not* order an ice-cream cone when within five miles of the University of Wisconsin-Madison campus. (For the record, my favorite flavor is one you cannot easily find elsewhere: Orange Custard Chocolate Chip.) If it is summer, I will enjoy outside, at the Memorial Union Terrace, watching sailboats muscle past and paddleboarders wiggle along Lake Mendota. An icy winter day does not sway me, although I will eat inside instead, on a wood-carved bench at the Rathskeller.

Crafted since 1951 on campus, at the Babcock Dairy Plant—part of the university's Food Science Department and named after UW researcher Stephen Moulton Babcok (in 1890 he invented the first reliable butterfat content milk test)—the creamery (making both cheese and ice cream) operates a scoop shop inside Memorial Union as well as at the plant itself. Select from twenty-two ice-cream flavors at any time, including Blue Moon (in the brightest of blues, the flavor was invented in Milwaukee during the 1950s) and Union

Utopia (vanilla with swirls of fudge, peanut butter and caramel). In recent years, the roster has expanded to include those made with Greek frozen yogurt as well as sherbet (such as grapefruit mango). Having a party with some Badger alums? Pick up a three-gallon tub for dessert (pints, half gallons, and—of course—by-the-scoop options are also available).

The creamery also crafts cheese and hires student production interns who learn, under the tutelage of Master Cheesemaker Gary Grossen, to make award-winning cheese Wisconsin is already known for. Taste for yourself by shopping in the store, where twenty-three cheese varieties are sold. Observations of the plant are free . . . until you are given the opportunity to order ice cream, which most do not mind because who can turn down ice cream? These tours, referred to as "observations," must be scheduled at least a week in advance. If you want to see production, make sure you visit between 9 and 11 a.m. weekdays. The second-story observation deck is always open, and at no cost, with machines labeled from this view so that, even without a tour guide, you can understand the process.

Farms and Farm Stands

Anu Sky Farm
1228 Lake Kegonsa Rd., Stoughton; 608-669-3059; anuskyfarm.com
While offering CSA shares, and with a stand at the Stoughton Farmers Market, this family-owned farm—which debuted in 2014—also hosts farm tours by appointment and believes in a mantra that "no one should go to bed hungry because they do not have the means to feed their family." Grown on the farm, using sustainable methods, are vegetables, flowers, fruit, and herbs.

Beefnbeaks Farm
N8248 Highway O, Waterloo; 920-478-2045; beefnbeaks.com
The name says it all: this is a farm that raises chickens, hens, and cattle. After watching the documentary *Food, Inc.* in 2010, owners Kris (former art teacher) and Dan Paape (worked in computer

programming) decided to take back their 150-acre farm . . . but first they had to "evict" the farmer renting the land for his own crops (corn and soybeans). While there are official events—like medicinal-plant walks for all ages and a farm camp for kids—visits and tours are welcome. Animals are grass-fed, pasture-raised, and free-range.

Bike or Bus The Barns

csacoalition.org

Ever bicycled through the Dane County countryside and been curious about the stories of farms you pass, and what crops they produce? This thirty- to forty-mile (or sixty miles, you pick) ride benefits Partner Shares Program, connecting low-income families with organic farm-fresh vegetables. Each of the three or four farm stops curates its own experience, such as live music, crafts, drinks, farm tours, and chef-made food right out of the fields. Not into bicycling or have mobility issues? Take the Bus the Barns tour instead. When: late September

Breakfast on the Farm

danecountydairy.com

Hosted at a different Dane County dairy farm for one day every June, this breakfast event is organized by the Dane County Dairy Promotion Committee. For around $8 ($4 for kids ages three to eleven) you get cheesy scrambled eggs, sausage, pancakes, yogurt, cheese, milk, and coffee—plus another reminder it is a dairy farm: ice cream. Also at the event: kiddie tractor pull, face painting (cow spots), "cheese hole" toss, and live music, with an opportunity to mingle with the state's Fairest of the Fairs (crowned at county fairs).

Burr Oak Gardens

W5511 County Road B, Rio; 608-234-0674; burroakgardens.com

A farm stand at this certified-organic farm and greenhouse—named for its three-hundred-year-old burr-oak tree and in the same family since the 1980s—sells not only plants (vegetable and herb start-ers) but seed packs, just-harvested vegetables and hanging flower baskets. For up-to-date info about what vegetables are being sold that week, check out the "Veggie Forecaster" on Burr Oak Gardens' website. Strawberries are sold in season. The garden also makes its honey and dried beans available for sale.

Country Bumpkin Farm Market

E9745 County Road P, Wisconsin Dells; 608-254-2311;
countrybumpkinfarm.com

Fall is when this family-owned market—perfect for kids, with a corn maze, outdoor play village (includes a zip line and obstacle course), train rides and petting farm—gets busy, although you can visit any time during the season (April through October). In business since 1997, the farm offers you-pick days (for strawberries, raspberries, blueberries, and pumpkins) and a bakery. Plant lovers will want to shop at the greenhouse. Everything—whether it is an ornamental plant or something you eat—was grown sustainably or organically.

Creek Bed Country Farmacy

N2767 Mountford Rd., Poynette; 608-635-8798;
creekbedfarm.com

This three-hundred-acre farm's cute pun of a name hints at its owners' (Darrell and Julie Schoeneberg) belief that farms are medicinal. As the fifth generation to farm Darrell's great-great-grandfather's land, they have transformed it into agritourism, with you-pick strawberries and sugar-snap peas (June and July), a sunflower festival with live music (August), and a corn maze with a pumpkin patch (plus hot cider and haybale-theater shows) come fall. "If they get lost [in the corn maze], they can also use their smartphones to pinpoint their location," says Julie. Grass-fed beef is also sold.

Deep Rooted

E8975 E. Ridge Rd.; Westby; 608-386-6177;
deeprootedorganics.com

Want to learn how to nail a ground-planter arrangement or hanging basket? This certified-organic farm—which offers a market-share program of its vegetables to La Crosse locals, plus a cut-flower subscription that sells out—will school you in a series of spring-planting

workshops. Every Labor Day is a celebration of tomatoes at the Tasty Tomato Festival, including the ones they grow. Get a jump start on your own garden by picking up certified-organic vegetable and herb seedlings; or annual or perennial flowers.

Deutsch Family Farm

N50666 Hogstad Rd., Osseo; 715-597-1815; deutschfamilyfarm.com
At this 160-acre hog, chicken, and cattle farm, which debuted in 2010, you can buy a half or whole hog—or bacon, brats, pork chops and ribeye steaks—at its pop-up markets or through local retailers, a local egg-roll food truck, and the Eau Claire Downtown Farmers Market. Hams are available around Easter. Partnering with K Point Brewing in Eau Claire, events include Bacon & Brews and Rib Fest. Deutsch Family Farm's Jersey cows' milk is used for Westby Cooperative's sour cream, butter, and cottage cheese.

Door Creek Orchard

3252 Vilas Rd., Cottage Grove; 608-838-4762;
doorcreekorchard.com

Growing "ninety-ish varieties of apples, several varieties of grapes
and pears, quince, American Persimmon, Paw Paw, and plums,"
says Liz Griffith, daughter of orchard co-owners Tom and Gretchen
Griffith, is the focus of this orchard with an on-store farm. Its eighty
acres date back to the early 1800s as a Norwegian-American dairy
homestead. Today the Griffiths reside in the 1860 farmhouse, since
opening the orchard in 1984.

August and September are when fruit matures for you-pick
opportunities (apples and grapes only) or pre-picked bags. Heard
of a Queen's Apple? It is a deconstructed caramel apple, and just
one of the items you can pick up in the store. Honey, fruit, and fresh
cider are also sold. Knitters and crocheters love yarn culled from
the farm's Black Welsh Mountain Sheep, who graze at pasture for
most of the year. In fact, the Griffiths even developed their own
sheep breed, only found at this orchard: Chocolate Welsh Mountain
Sheep. Most everything the family grows is sold at the farm, outside
of a few restaurant partners in the Madison area—chefs who turn
their bounty into decadent dishes on the menu.

"We sometimes have food trucks serving different foods on
select weekends," says Griffith. "We also have an 'Heirloom Apple
Dinner' on our farm in early November that celebrates our later season heirloom varieties with a chef-created meal." Only forty tickets
to the five-course dinner are sold, divided between two seatings.
The dinner kicks off with a welcome cocktail (such as the orchard's
cider with sparkling wine and bitters) and winds down into dessert
(in a past year, this meant an apple tart with caramel and brown
butter ice cream, or a cheese plate with apple brandy)

Being sustainable is an important mantra for the Griffiths and
drives everything they do. This translates to ecologically advanced
integrated pest management and low-input farming. The idea is to
grow crops with minimal impact—and harm—to the environment.
You also won't find a hayride or other attractions typically linked to a
fruit orchard. "We are not an 'agri-entertainment' farm," says Griffith,

"so we do not have many formal 'experiences' to offer. Our focus is on our fruit and our land."

Dreamy 280 Farm Fresh Meats

2792 Cave of the Mounds Rd., Blue Mounds; 608-437-8074; dreamy280.com

Dennis and Lisa Schlimgen named their 280-acre Angus and Short-horn cattle farm in honor of their dream career as farmers. Both grew up on local farms and live in the property's 1914 farmhouse. Specialty beef cuts are sold at its store (plus chicken meat, eggs, pork, canned goods, and crafts from other farms), and they organize beef cook-offs. "It is important that we implement sustainable beef-raising practices in order to help produce safe, wholesome, afford-able food while using fewer natural resources," says Lisa.

Driftless Organics

52450 McManus Rd., Soldiers Grove; 608-624-3735; driftlessorganics.com

If you shop at a natural-foods store in Wisconsin, chances are you have seen this farm's organic sunflower oil on the shelves. It is considered an innovative, somewhat new cooking oil, cold expeller pressed sunflowers. Owned by Josh and Noah Engel, who sell their vegetables at Dane County Farmers Market, the farm's current chapter rose out of its organic-dairy farm roots in 1993, when the boys were young. Each September is a fall harvest party on the farm.

Five Green Acres

605 McMillan St., Poynette; 608-335-9444; fivegreenacres.com; and atmosphericfloral.com

"Chickens are the gateway drug, as many know, and soon led to a flock of fiber sheep, more poultry and fowl, pastured pigs, and a dairy cow," jokes co-owner (with husband Andrew) Mary Jo Borchardt. In 2016 they reimagined their five acres into a flower farm, launching Atmospheric Floral (full-service floral design) in 2020. Visitors can drop by for workshops, open houses, and a farm-to-table dinner in the flower fields. Bunches, market bouquets, or bulk buckets (for DIY design) are sold through the website and to local florists and designers. Succulents are local delivery only.

Fleming Orchards
46054 WI-171, Gays Mills; 608-735-4625

James and Ruth Fleming operate this orchard, which dates to the 1930s, and also sell at the Dane County Farmers Market. By early September, pick-your-own apples, and a full-on bakery (items include apple pies) debut at the farmstead, transitioning into all things pumpkin shortly after. Caramel apples satisfy sweet-tooth visitors and items like honey, jams, spices, potatoes, onions, and gourds help stock your pantry for fall. Goats on the property are continually in search of petting opportunities.

Four Elements Herbals
111 E. Walnut St., North Freedom; 608-522-4492; fourelementsherbals.com

I was in my twenties and a wide-eyed, new resident to Madison when I stumbled upon Four Elements Herbals' aromatic and exceptionally calming body-care products at the Willy Street Co-Op. I was stunned to learn they are made in North Freedom, only an hour's drive from the store. Even better? Herbs grown on this farm within the Baraboo Bluffs are folded in. This may mean peppermint and thyme in a cream to moisturize your feet, or calendula (also known as pot marigolds) paired with orange flavor in a balm to soothe cracked lips.

I now shop at Outpost Natural Foods Co-op in Milwaukee, where I have witnessed the the Four Elements Herbals product line expand. It includes herbal teas, lip balms, salves, body oils, outdoor body sprays, soap bars, herbal tinctures and skincare creams

Four Elements Herbals was started in 1987 by Jane Hawley Stevens, armed with a horticulture degree from UW-Madison. Designs for herb gardens captivated her and she dug deeper into this passion by using herbs for cooking and natural remedies. Ingredients that go into the brand's products are harvested from Stevens's 130-acre certified-organic farm, with the help of her husband David.

Each summer Stevens opens the farm and apothecary for weekend-long workshops, sharing her expertise while participants work alongside her to collect herbs from the garden, and process them (drying and storage) into products for take-home use. There is also an apothecary-style shop in North Freedom, inside a former bank branch. With three decades of business behind her, Stevens has managed to get Four Elements Herbals products in stores nationwide, not just in Wisconsin. You can be shopping at a natural-foods store in Anchorage, Alaska, or just south of New York City's Central Park, and find Four Elements Herbals products. They are also in select Whole Foods Market stores nationwide. In 2020 the brand achieved another recognition: MOSES (Midwest Organic & Sustainable Education Services) named the couple Organic Farmers of the Year.

Gander Family Farms
9127 S. Landers Rd., Beloit; 608-364-4767; ganderfamilyfarms.com
This beef farm operates a store (first and third Saturdays of the month from 10 a.m. to noon) and farm stand (summers). Its beef boxes offer half or full shares of roasts, steaks, ground beef, and specialty cuts. Included recipes help drum up a delicious dinner and customers also like the boxes because they do not have to commit

to an entire steer, but can still support a local meat farm. Themed boxes of beef include a Bone Broth Kit and a Grilling Kit (four steaks and four pounds of ground chuck).

Green Pastures Cattle Co.

S8857 Highway C, Plain; 608-432-8185; greenpasturescattleco.com
Owners Courtney and Aaron Feigl give new meaning to using the entire animal: Aaron, a third-generation farmer, uses his taxidermy talent for skulls and Texas Longhorn Mounts, sold in the farm's shop, along with butcher meats (from the Texas Longhorns they raise), chicken eggs (also from the farm), canvas artwork, T-shirts, hats, and mugs.

Hilltop Community Farm
S850 Heidrich Rd., La Valle; 608-257-6729;
hilltopcommunityfarm.com

This sixty-acre farm is so gorgeous it serves as a wedding site. Erin Schneider and Rob McClure grow vegetables, fruits, and flowers that ultimately goes to CSA members—and often with add-ons like pickles, preserves and salsas—as well as local restaurants. The couple launched the farm in 1993 and taps into innovations like solar energy and a true lack of machinery.

"We grow a diverse variety of organic produce, fruits, and flowers. Fruit includes serviceberries, currants, elderberries, aronia berries, raspberries, pears, apples, and hardy kiwi, and in a good year cherries and plums," says Schneider. "Flowers encompass over eighty species, primarily native perennial plants that work well as cut flowers, alongside annual blooms like zinnias, sunflowers, lisianthus, and dahlias."

"Our flower menu includes custom bouquets, buckets of blooms, and full-service floral design and installs for weddings and events," says Erin. Orders are taken online, by phone, or through the farm store. Weddings and other events have also hired the farm to create bouquets and arrangements.

Does creating a floral arrangement scare you? Flower-arrangement classes are hosted at the farm. These "Brunch n Blooms" events coach in just that, while also serving a yummy brunch with items like quiches, pickles, and breads that fold in what is grown at the farm.

"We are grateful for the land we steward. The farm's sixty acres include sunsets that take your breath away," says Erin. "An oak savannah since the last glaciation, later part of the Ho Chunk Nation before European occupation and settlement, and a dairy operation when it was abandoned in the mid-1940's, Hilltop's house and outbuildings paused while the land was farmed by nearby residents until 1972, when the parcel was purchased by Rob's parents Donald and Anne McClure of Winnetka, Illinois."

"The early-1900's house was upgraded with electricity and running water then restored to a circa-1920s rural residence that we dwell in today," says Erin.

Homestead Wisconsin

7741 State Highway 69, Belleville; homesteadwisconsin.com; and awellwornstory.com

Neither Brit nor Matt McCoy had grown up on a farm or even spent a lot of time on farms. But they knew, without question, they wanted to be farmers. Raising grass-fed Scottish Highland cattle and North Country Cheviot sheep (as lamb meat and for their fibers) is just the start of their success. They have also dipped into flowers, offering monthly subscriptions, and providing couples with floral art on their wedding day.

They are so passionate about what they have done to the land that visitors are encouraged. This may mean staying overnight, taking a tour with Brit and Matt (either solo or with a group, so long as you call in advance), or signing up for a floral-design workshop. Flower on the Farm workshops are held in the evening and serve as an intimate opportunity to get to know Brit while sipping wine and relaxing on the farm. The workshop fee includes greens, floral snips, and flowers; plus, a vase to design and display your bouquet in.

"Historically our farm was a sheep farm, as much of our land features steep slopes that are perfect for grazing sheep," says Brit. "In the late 1970s, with a change of ownership, it became a fish farm,

but after a few years they closed the doors on that operation and until 2017, when we purchased it, it sat unmanaged. Our neighbors and the locals still lovingly refer to it as 'the fish farm,' as that is a unique farming venture in Wisconsin."

Their meats, which in addition to grass fed are humanely raised, can be bought online (a handy YouTube video on the farm's site walks you through the process), with free delivery within twenty miles. The farm also offers membership in their Grazer Club, all about packaged meats.

Beyond meat and flowers, Brit collaborates with her sister-in-law, Morgann McCoy, on a line of leather goods (leather totes, messenger bags, belt bags, wallets, clutches, cuff bracelet, and belts), that trace back to their cattle hides. The name of that business is A Well Worn Story. Morgann studied at the London College of Fashion and earned a degree in Apparel Management from Iowa State University.

Links Greenhouse & Farm
N9905 Link Rd., Portage; 608-742-6758

It is rare for a greenhouse to be open year-round, but that is what keeps Links interesting, with an activity for each season, such as cut-your-own trees just before Christmas (tack on poinsettias or wreaths) and pumpkins for sale in autumn. Since 1967, Links has been family owned. Local gardeners know to pick up plants (herbs, annuals/perennials, vegetables, and pond plants) and trees beginning in spring, and—come late spring—farm-fresh vegetables. You-pick asparagus, also available.

Lutz Family Farm
1477 County Road T, Marshall; 608-655-3664; lutzfamilyfarm.com

Located just a half-hour drive from downtown Madison, this family-owned farm offers you-pick strawberries (middle to late June) and sweet corn (late July through August). In a rush? Pre-picked strawberries and corn—along with blueberries—are sold at the farm. The family posts daily updates about what is in season—and when you can come out to pick—on its website and Facebook page.

Mad Lizzie's Flower Farm

5700 Otto Kerl Rd., Cross Plains; 608-669-8958; madlizzies.com
This funky name (named after their first daughter, Madeline, named for Wisconsin's Madeline Island) hints at the fresh spin Katy and Dale's sustainable flower farm—tucked into Cross Plains, just west of Madison—puts on flowers, including fun events like you-pick sunflower "happy hours" on Fridays and Buckets O'Blooms (you design the bouquets with the fifty to seventy-five stems in the bucket). Weekly, bi-weekly, or monthly subscription series of fresh-cut floral bouquets are joined by a roadside stand selling potted herbs or flowers, and wrapped flower bouquets.

Mastodon Valley Farm

Viola; 608-843-6467; mastodonvalleyfarm.com
This farm raises grass-fed cattle, hogs, Freedom Ranger chickens, and lamb for meat. Rambouillet Sheep wool is spun into yarn and sheepskin rugs. A five-day regenerative farm design course every August is for aspiring farmers who want to do good by the land. Peter Allen's farming career came out of a passion for restoring Oak Savannah ecosystems and realizing food should come from them—not factory farms or unnatural methods. Purchase meats through the online store—many say they detect mint in the lamb, as a result of sheeps' nibbling on this herb at pasture.

New Life Lavender & Cherry Farm

E10766 County Road W, Baraboo; 608-477-4023; newlifelavender.com
Two summers ago, while staying at Sundara Inn & Spa in Wisconsin Dells—where the entire time I kept asking my sister-in-law, "Are you sure we're not in Sedona or Scottsdale?" because the property is that luxurious—we dropped by this new lavender farm that had just opened its doors. Grown on the property are cherries, herbs (for use in their teas and drinks), and twelve thousand lavender plants, which are in bloom mid-June through mid-July.

From the parking lot I immediately knew that Laura and Aron McReynolds's forty-acre farm would be different, a slice of Provence adjacent to the water-parks capital of the world.

"We bought a forty-acre farm property that had been abandoned," says Laura. "In a previous life, it had been a working dairy farm and homestead, and known as a place of neighborhood gatherings in years past. We find it fitting that the abandoned property is now restored not only for the beauty of the land, but as a place for family and friends to come, gather, and relax in our lavender fields."

There is lots to do on a visit, including wagon ride tours around the property. "The tour includes aspects of growing lavender, beekeeping, the chance to feed and pet our sheep, and hear a few funny fish jokes," says Laura. You-pick lavender days, wreath making, and distillation demos are in July.

The retail store (open daily except for Sunday and Monday) illustrates all that can be made from lavender. As we shopped, my arms became heavy with jars of lavender/peach salsa and bags of loose-leaf lavender tea. I was already eyeing the prospect of an iced lavender latte. Next time, I will take Laura's advice and order the most popular drink: blueberry lavender shake. Cherry lavender pie is baked fresh daily and lavender ice cream is crafted on site.

"There are many opportunities to see beautiful flowers all around our farm, sit on the back or side patios to enjoy the views, or sit in our Secret Garden surrounded by our one-hundred-year-old wall," says Laura, who also hand-sews all fabric items, such as dryer sheets and sachets.

Peck's Farm Market

6446 US Highway 14, Arena; 608-588-7907; pecksfarmmarketeast.com

Attracting Shakespeare lovers en route to a performance at the American Players Theatre, or Frank Lloyd Wright fans off to see his Taliesin homestead, this market is also thrity-five miles west of Madison. It has been in business since 1889, always functioning as a produce stand. You can also find Amish candy, gifts, and crafts and—if you like—pet a deer or hop into a horse-drawn wagon.

Rainbow Fleece Farm

W7181 Hustad Valley Rd., New Glarus; 608-527-5311; rainbowfleecefarm.com

Local fiber artists love Patty Reedy's farm for its fiber shop, which sells supplies and patterns (plus yarn spun from its sheep) for felting, needle felting, spinning, knitting, and crocheting. Do not be scared if you do not work with yarn, though: each April's sheep shearing, "Farm to Table Turkey Trot" (coinciding with turkey pick-ups in late November), and "Holiday Yarn on the Farm" (selling sheepskins, knit hats and socks, and felted laundry balls for the holidays) are open to anyone. Chemical-free chicken meat and farm-raised lamb meat is also sold.

Raleigh's Hillside Farm

901 N. Marsh Rd., Brodhead; raleighshillsidefarm.com

"The Leek and the Carrot" blogger Lauren Rudersdorf (whose recipes and food photos will make you drool) and her husband Kyle met while both attending the College of Agriculture & Life Sciences at the University of Wisconsin-Madison. In 2013, they opened their organic vegetable farm, leasing four acres on Lauren's family's land, and tapping into Kyle's farmhand experience at Tipi Produce. CSA members receive shares of what is grown. Keep an eye on the farm's

website and Facebook page for news of tours, wellness workshops, and a harvest party in September.

Richland County Dairy Breakfast
fair.co.richland.wi.us
Always held in mid-June, this traditional breakfast moves to a new dairy farm each year. In 2021, the 40th anniversary will be marked, hosted at Hatfield Dairy Farms in Viola, a farm with 200 cows and owned by the Hatfields, who have been farming together since the 1980s. The breakfast's goal continues to be highlighting all the many ways you can whip up a delicious breakfast using items sourced from a Wisconsin dairy farm...by tasting the products yourself.

Riemer Family Farm
W2252 Riemer Rd., Brodhead; 608-897-6295; riemerfamilyfarm.com
Pick up your meat, embark on a self-guided tour, or drop by on a field day (educational tour). Jen and Bryce Reimer—who fled Chicago's 'burbs in 2010 to take back the family farm and raise their three kids—are the third-generation owners of this 280-acre ranch. Meats raised and sold include pastured pork, chicken, and turkey, and grass-fed beef and lamb. Eggs are available, too, and meat can be shipped.

© Riemer Family Farm

The farm's a stop on the annual Soil Sisters tour in early August, when female-owned farms like this one are highlighted.

Roots Chocolates
608-963-5319; rootschocolates.com

While you are more likely to find a chocolate farm in Mexico or Belize than you are in Wisconsin, those destinations do not have the kind of soil Upper Midwest farmers love. This allows fourth-generation farmer (and Chief Chocolate Officer) Lisa Nelson—farming on her family's land—to fold in fruits, herbs, and vegetables produced on the farm into hand-crafted chocolate bars and truffles.

Her chocolates' names and inspirations are often derived from local culture. For example, "Beer Naked" marries dark chocolate with Wisconsin craft beer and pretzel bits, and "Ol' Fashion" is a twist on an Old-Fashioned cocktail incorporating muddled juice from the farm's cherry tree. One of her chocolates—Lapsong Souchong (Lapsong Souchong tea infused with dark chocolate)—placed in the International Chocolate Awards in 2017.

The farm's squash even makes its way into a chocolate, with "Sass-Squatch," joining Wisconsin maple syrup, ginger, and honey. "Much of what I grow I do with the intention of testing it in my choc-olates," says Nelson, who studied to be a chocolatier in Vancouver and Kansas before opening in 2010. "Examples are aronia, apples, pears, quince, cherry, mint, basil, elderberry, and garlic." Blending, cooking, cutting, and packaging takes place on the farm, in a com-mercial kitchen.

And although chocolates are the current industry, she knows a lot about the farm's history. "It was originally one hundred acres," Nelson says, "but is now about twenty. There have been many inter-esting animals here, but my favorite was the herd of bison (about twenty-five of them) that were raised for meat."

Tours of the farm are offered on Wednesday afternoons and chocolates can be picked up with advance notice. "I offer walking tours around the farm and introduce folks to what I grow and how I grow it, along with a small sampling at the end of the tour," says Nel-son. She also sells at the Dane County Farmers Market, Madison's

Northside Farmers Market, Sundara Spa in Wisconsin Dells, and Metcalfe's Market's Hilldale location in Madison.

Sassy Cow Creamery

W4192 Bristol Rd., Columbus; 608-837-7766; sassycowcreamery.com

Dairy farms are a dime a dozen in Wisconsin but not all have built a brand found in grocery stores. Nor do they make ice cream. Since its 2008 debut—the brainchild of brothers James and Robert Baerwolf and their families—Sassy Cow Creamery's distribution of its milk and 60-some ice-cream flavors has expanded to around 100 stores and 150 restaurants, in Wisconsin, Illinois, and Minnesota.

On a hot day last summer, the line to order a scoop at the organic creamery's cafe was long, but you need time to decide among twenty-six flavors. Also sold at the cafe are caprese sandwiches folding in Crave Brothers Farmstead mozzarella, and decadent treats like a strawberry-cheesecake ice-cream sundae. Locals can score farm-fresh groceries from not only the creamery (such as its milk and ice cream) but also Fischer Family Farms' meat, artisan cheese from Wisconsin creameries, and other snacks made around the state.

The education piece is a tour of the farmstead, just a five-minute drive from the creamery. It begins with participants seated on milk crates to view a short video about the farm and creamery, followed by the guide's dispensing of facts about cows—including what they eat, the differences between many breeds, and how often they are milked (4 a.m. and 4 p.m.). Then, you view from behind a glass wall the milking parlor's "merry go round," which the resident cows love, followed by a walk alongside the pasture and into the barn. Feeding and petting of these cows is encouraged, although many are shy. Then, at the end of the tour, you get to grab a small milk bottle out of a refrigerator—chocolate, strawberry, or "plain." Back at the store, bottling and the ice-cream making process are viewable through glass windows (when happening).

A new offering is outdoor yoga and mindfulness meditation taught by Tanatnan Chaipang of Tanatnan Yoga Monday and Thursday evenings. Families flock to the store for the twist on typical

ICE CREAM
1 HALF GALLON (892 ML)

Sassy Cow
CREAMERY
WISCONSIN DAIRY

ICE CREAM

DARK CHERRY CHOCOLATE

Sassy Cow
CREAMERY
WISCONSIN DAIRY

ICE CREAM
1 QUART (945ML)

WORLD DAIRY EXPO · **Dark Cherry Chocolate**
Cherry ice cream with cherries and chocolate flakes

ARCTIC MINT FUDGE
ARCTIC MINT FUDGE

MINT FUDGE

1st PLACE

Sassy Cow
CREAMERY

ICE CREAM
1 HALF GALLON (892 ML)

Sassy Cow
CREAMERY

ICE CREAM
1 QUART (945ML)
Made in Wisconsin

ARCTIC FUDGE
MINT FUDGE
ARCTIC MINT

Sassy Cow
CREAMERY

ICE CREAM

Sassy Cow
CREAMERY

ICE CREAM
1 QUART (945ML)

Sassy Cow
CREAMERY

ICE CREAM

Arctic Mint Fudge
Mint ice cream with fudge pieces

playground equipment—a tractor, not a jungle gym—and that there are baby goats and calves to see.

Sinsinawa Mound Collaborative Farm

sinsinawa.org

Acting as an incubator for aspiring farmers who only lack access to land and education, this four-acre, certified-organic farm was launched by Dominican Catholic Sisters in 2017, in tiny Sinsinawa. Farmers with less than ten years of experience receive a plot, cold storage, irrigation, access to a washing and packing shed, equipment, and classes and workshops. The on-site bakery has been going strong since the 1960s and sells to businesses in northeast Iowa and southwest Wisconsin.

Seven Seeds Farm

5079 County Road Z, Spring Green; 608-225-2892; sevenseedsorganicfarm.com

This multi-generational farm continues what the Mickelsons from Norway began . . . only it is no longer a dairy farm. They pasture-raise chicken and pigs, and also grass-fed beef, on 170 acres. They even grow their own feed. Shop for meat—on the menu at restaurants like Graze in Madison, and summer sausage at Willy Street Co-op—at the farm's store (Wednesday from 3 to 6 p.m. and Saturday from 11 a.m. to 3 p.m.). With a little notice, visitors can tour the land and see the animals.

Ski-Hi Fruit Farm

E11219A Ski Hi Rd. Ste. A, Baraboo; 608-356-3695; skihifruitfarm.com

The season at this fifty-acre farm—dating back to 1907—kicks off in late August when apples are ready and cider is pressed. Couples can even get married in the farm's orchard, using apple bins and vintage tractors as props. Cider donuts and Grandma Olga's pies are baked daily and, on Fridays, apple fritters. Kids love to pet the barnyard animals and wander through the family's 1862 log cabin, a Sauk County historic site. Or pick up apple salsa and a bright-red hoodie with the farm's logo.

Sprouting Acres Farm

1746 Highway 73, Cambridge; 608-469-2319; sproutingacres.com

In 2019, this certified-organic, twenty-acre vegetable farm began hosting pizza nights (first and third Sundays) with live music. This new generation of farmers—Andy Watson, who calls himself head farmer/manager/instructor and earned a soil-science degree from UW-Stout; and his wife Kelly Bratt—is reinventing land his parents farmed as a tree nursery since the 1970s. "My parents still live on the farm and help out watching our children, around the farm, and on pizza nights," says Watson. Two-hour-long cooking classes in a barn built in 2018 include recipes and a beverage while you dice, slice, and learn, looping back to Watson's past life cooking in Madison restaurants like Harvest.

St. Francis Hermitage

14841 Turfan Rd., Gays Mills; 608-735-4015; stfrancishermitage.org

There are certain things you expect in the Wisconsin countryside. A fine-dining French eatery is not one of them. And yet that is where the Fraternite Notre Dame friars and nuns in this town work, using produce grown on their farm, especially in the Hermitage and St. Francis Orchard salads. The restaurant's open to the public Saturdays and Sundays for breakfast and dinner, offering menu items like Burgundian escargots and Antilles-style Cristophene au Gratin.

Sunborn Gardens

Mount Horeb; sunborngardens.com

In business since 1975, this farm owned by Hans and Lisa Larsen (the farm was his mother's) brightens up customers' days with bouquets of pre-ordered cut flowers (picked up on site or delivered to Madison porches), summertime subscriptions, or a bucket of blooms you arrange (Fridays). Wedding couples love the flowers, too. Workshops coach in stunning arrangements, such as one in hand-held lisianthus bouquets. Dried flowers are sold for a longer shelf life and come November and December, holiday-themed arrangements and wreaths are available. Among the flowers grown are sunflowers, zinnias, tulips, dahlias, snapdragons, Sweet William, and delphiniums.

Suncrest Gardens Farm

S2257 Yaeger Valley Rd., Cochrane; 608-626-2122;
suncrestgardensfarm.com

It is pizza night on Fridays and Saturdays—even on select dates in
the winter months when it moves into the barn for a cozy vibe. Beer
and wine can be tacked onto your wood-fired pizza order and there
are plenty of gluten-free or vegetarian options (for pizza). Tours of
the property can be arranged by appointment, guided by owner
Heather Secrist. She lives on property with her husband—who
manages his family's dairy farm nearby—and two young sons. There
is a harvest dinner every November.

Sylvan Meadows Farm

E8303 County Road SS, Viroqua; 608-606-2369

John and Virginia Goeke established the sixty-five-acre farm in 1994
and enjoy the farm lifestyle with their son, Sylvan, for whom the
farm was named. At their stand and store, they sell grass-fed beef
(heritage breeds and Belted Galloway) and lamb; pastured poultry
and heritage pork; plus, organically raised fruits, heirloom vegetables, and flowers—and even kombucha, laundry and dish soap, and
hand-spun yarn. A festival every September is complemented by
workshops in soapmaking, cooking and preserving, and wool crafts.

Ten Eyck Orchard

W968 State Roads 11 and 81, Brodhead; teneyckorchard.com

Owned by sixth-generation, father-and-son Rob and Drew Ten Eyck,
this three-hundred-acre farm sells its fruit, honey, apple butter
and pastries, plus squash and pumpkins. A corn maze and horse
rides debut each fall. "It was a dairy farm, which is why we have a
round barn, until World War II when many circumstances caused
him [my grandfather] to sell the cows," says Drew. "He had a couple
large standard apple orchards that my father slowly converted to
high-density dwarf plantings. Now we grow sixty cultivars as well as
plums and pears."

Rooted

Badger Rock Urban Farm, 501 E. Badger Rd., Madison;
Try Farm, 502 Troy Dr., Madison; rootedwi.org

Two certified-organic vegetable farms—both in urban pockets of
Madison—are under Rooted's non-profit umbrella: Badger Rock
Urban Farm and Troy Farm. A major goal is to provide healthy food
to locals who may not otherwise have access, while at the same
time fostering appreciation in pursuing a career as a farmer. In the
Urban Farmer Training Program, skills like community organizing and
growing organic on a small scale are taught.

Badger Rock Urban Farm's South Side location, at Badger Rock
Neighborhood Center, teaches students at Badger Rock Middle
School how to tinker with the soil, growing fruits and vegetables
from seed using organic growing practices. Under the tutelage of
Rooted staff, they start seeds in a heated greenhouse during winter,
then transition those plants to hoop houses and the outdoor garden
come spring. The work does not stop there: Former students are

retained during the summer months in exchange for a small stipend as they see the plants through to fruition. When you purchase these plants at its farm stand, proceeds fund the school's program. Workshops and classes in urban farming are offered as well. On the second Friday of the month, free CommUNITY dinners illustrate the decadence that can be crafted out of the harvest, at the hands of Roots culinary-arts manager chef Kipp Thomas.

Dating back to 2001, Troy Farm is one of Madison's oldest urban gardens, tucked into the North Side. Farm members subscribe to regular allotments of farm-fresh vegetables. A Thursday-night farm stand is joined by a pay-what-you-can community meal and live music. You can also support the program by buying produce at its Northside Farmers Market stand, on Sunday mornings. Each spring a vegetable-plant sale draws many locals out to the farm, too.

Both farms also offer you-pick flower beds, so long as you are a member of the farm. And if you cannot get out to the farms, but still want to lend support, sprouts (lentil mixes, mung beans, broccoli, peas, and alfalfa and clover mixes) are sold year-round at Willy Street Co-op locations, the East Side Hy-Vee and Metcalfe's Market's Hilldale location.

West Star Organics

2555 West Star Rd., Cottage Grove; 608-347-4346;
weststarorganics.com

Recently branching into hemp products (for sale in its online CBD
store), this seasoned grower of certified-organic plants (starters of
herb, vegetables, and flowers, available April to June) operates on
a forty-acre farm setting. Organic gardening products important
to cultivating healthy soil (like compost and growing mix) are sold
year-round. Visitors are invited out to not only shop but also enjoy a
wooded walking trail.

Winghaven Pizza Farm

N18057 Grover Ln., Galesville; 608-519-4228;
winghavenpizzafarm.com

The name says it all: Since 2017, people have come here for stone-
fired pizza. Fifth-generation owners Rob Grover and Greg Roskos—
both longtime friends; Rob's family has farmed the land since the
1850s—also honor the former Winghaven Orchards by retaining
their name. Pizzas are paired with local craft beer and wine, live

inghaven Pizza Farm

music, and outdoor play equipment for the kids. "What we do not grow on the farm, we are proud to locally source from around our area," says Grover.

Winterfell Acres

W1912 Mortensen Rd., Brooklyn; 608-628-7504; winterfellacres.com

After reading *The Dirty Life: A Memoir of Farming, Food, and Love* by Kristin Kimball, and interning at a Franciscan Sister's two-acre farmette, Beth Wright yearned to be a career farmer. She and her husband—and young daughter, who was born in the farmhouse— harvest fruits, vegetables, herbs, and flowers. Each August, the farm—which debuted in 2014—is on the Soil Sisters tour of female-owned farms, offering workshops in topics like how to make offerings of gratitude for your land and home, with take-home offerings that include bay leaves, ground rainbow corn, and tobacco.

Farm Stays

The Birdhouse Inn B&B

6901 High Point Rd., Arena; 608-588-6272; thebirdhouseinn.com

Emily Martorano (a Harvard grad who studied anthropology and was raised in a Chicago suburb) and Ana Maria Vascan met in a Madison book club in 2017. Their next chapter? Co-owning and managing this super cute three-suite B&B with an ecological bent two years later, and on the same fifty-five-acre property as Hazel Hill Farm, where Emily farms mushrooms, pork, eggs, and garlic. It is surrounded by mature forest: oak, hickory, birch, cedar, and pine trees. Vascan is a researcher in sleep neuroscience at the University of Wisconsin School of Medicine and studied architecture at the University of Wisconsin-Milwaukee, as well as art at the Milwaukee Institute of Art & Design. Much of the B&B's aesthetic is all her.

"Humphrey 'Bogey' Bogart, the guardian dog [a Maremma, a typical guardian dog breed from Italy], loves meeting new people," says Vascan. "He lives in the pig pasture and is very excited when guests come down to visit him, his pig friends and the 'chicken valley' where our flock lives in a 1970s trailer. The several-hundred shiitake logs zipper up the hillside near the driveway and guests are welcome to walk around and explore. If guests are very interested, they can accompany Emily on her (very early) morning or evening chores."

Less than two miles from the B&B is a reason many, pardon the pun, flock to The Birdhouse Inn & B&B each summer and fall, from as far away as Chicago and the Twin Cities. The American Players Theatre, one of the country's finest examples of an outdoor theater, has produced Shakespeare, comedies, and classical plays since 1980. Another draw is Frank Lloyd Wright's nearby former home-stead, Taliesin, which offers guided tours and pop-up events like chef dinners and classical-music nights.

"Rates include a 'slow food' farm-to-table breakfast," says Vas-can, which includes French Press coffee and loose-leaf teas, plus as much local produce as possible. Three rooms are suite-style with two bedrooms and a private bath each—The Treehouse, The Solar-ium Suite, and The Eyrie.

Campo Di Bella Winery

10229 Sharp Rd., Mt. Horeb; 608-320-9287; campodibella.com

With a twenty-acre setting that includes heritage lamb, a vineyard, a farm dog named Stella, and overnight stays—plus three-course dinners on Fridays and Saturdays—this is not your average Wisconsin winery. Open since 2014, it is what you would find in Italy, as the name suggests. Book the solar-powered farmhouse's private suite through Airbnb, flaunting gorgeous sunsets from its second floor. In addition to the Frontenac and Petite Pearl vineyard, the owners (of Italian heritage) also harvest fruit and vegetables. Their Vermouth is made from Wisconsin grapes, Michigan brandy, and mulled Penzeys spices.

Circle M Farm

**1784 County Road H, Blanchardville; 608-636-4652;
circlemfarm.com**

Kriss Marion is like your cool big sister. In 2005, she and her husband Shannon abandoned bustling-but-stressful lives in Chicago to
raise their kids on a Wisconsin farm. They found utopia in this small
town forty miles southwest of Madison.

In her cozy restored 130-year-old farmhouse and studio with an
Anthropologie-meets-Kate-Spade, 1940s vintage vibe (think boho-
chic décor and cute printed fabrics), Marion hosts girlfriend retreats
and team-building activities for local companies like Lands' End.
One day a class may mean spinning (and then dying) fibers culled
from bleating sheep just outside the window while on another it may
be gardening. Marion also teaches fiber-art and felting classes at
Shake Rag Alley in Mineral Point and Folklore Village in Dodgeville.

In recent years Marion—who calls herself a baker, farmer. and
"changemaker"—expanded into farm stays, including a 1968 Holiday Trav'ler decked out with lawn games. Guests can also bunk in

two farmhouse guestrooms where vintage handmade quilts and a shabby-chic aesthetic await. A woodstove downstairs is encircled by three couches. Access to a wood-fired hot tub, plus loaner bicycles and kayaks, is also available to guests. Because the Pecatonica River borders the property, kayaking is a few yards away, and trout fishing is nearby (pack your waders). Marion's library of homesteading books may keep you up late into the night planning your own farm dream. Coming for the day to take a class and play on the farm (whether it is you-pick in the fruit fields or helping with farm chores) is an option popular with those driving in from Madison or Milwaukee. Quite a few writers, photographers, and artists have come for a DIY creative retreat.

As a certified-organic vegetable farmer Marion's active with the Soil Sisters, a group of female farmers. Each August, for three days, Soil Sisters' self-guided (but curated) tour includes a stop at Circle M.

Breakfast is when Marion truly shines, dishing out homesteading tips at the waffle bar or as she brews locally roasted coffee. Many items are sourced from her farm, including cream from the goats, eggs from the henhouse, and sausage from the hogs.

Dorothy's Range
W8707 Sawmill Rd., Blanchardville; 608-444-1102; dorothysgrange.com
On this 50-acre farm—named for Dottie, the mom of one of the owners—that raises heritage pork are a small cottage and two rooms (in the main house) for overnight stays. Or you can book a tour to view the trout stream, wildlife pond, prairie and gardens. Owners Steve Fabos (an expert in restoring native prairie) and April Prusia (a seasoned farmer) love to teach others through workshops that touch on medicinal-plant identification, swine management, fermentation, and crafting root beer. Funds from celebrity chef Rick Bayless' Frontera Foundation restored the hog barn.

The Farmers Inn
E7830 Anderson Rd., Viroqua; 608-675-3553; thefarmersinn.net
Nestled in Wisconsin's most prolific boutique-farming region, this farm offers stays that include a tour, petting resident horses, and

gathering eggs from the chicken coop. Lodging (one party at a time, so it is totally private) is in a two-story cabin crafted from pine and includes a full kitchen so you can make meals using ingredients grown on site or at nearby farms. Sleep in the loft (two single beds) or bedroom (one queen bed).

Haymow Barn House at Prairie Creek Farm

6620 Johnson Dr., Ridgeway; 608-341-9196;
airbnb.com/h/prairiecreekfarmwi

Stay in the one-hundred-foot-long barn house on this 412-acre, sixth-generation working farm, where the Johnsons planted a vineyard in 2020. The barn house's sitting area is a former corncrib, and there are two queen beds and two bathrooms, plus a deck. "We actually have people message us, asking if the barn cats (Morrison, Sharpay, and Shaggy) will be at the farm during their stay," says Toni Johnson. "People are [also] fascinated with the cattle." Join the farm crew at feeding time or simply walk the grounds.

Inn Serendipity Bed & Breakfast and Farm

7843 County Road P, Browntown; 773-706-4150;
innserendipity.com

When John Ivanko and Lisa Kivirist left advertising-industry jobs in Chicago in the 1990s—in pursuit of a bucolic farm lifestyle in Wisconsin—they were among the first to launch an agri-tourism business. Today, you can stay the night at their two-bedroom, eco-friendly B&B dubbed Inn Serendipity. It is set on five and a half acres just outside of the town of Monroe. While they quip that they are known as that "interesting couple on County P," there is no place they would rather be. They are now raising their teenage son on the farm.

Inviting people onto their homestead to spend the night, work for a day, take a walking/tasting tour, pop in for "Pizza on the Farm," take a renewable-energy workshop, or observe organic gardening and vegetable farming quickly became their passion as they dived into their new lives as farmers. (I even helped install their strawbale farm one fall day, somewhere around 2001!)

Kivirist established a network of female farmers around the state called Soil Sisters. (Learn more by reading her book, *Soil Sisters: A Toolkit for Women Farmers*.) Now they can share resources and knowledge, encouraging each other along the way. During the first weekend in August, a handful of female-owned farms in the Soil Sisters network—including Inn Serendipity—open their doors to the public. Kivirist also launched In Her Boots, a project through MOSES (Midwest Organic Sustainable Education Service), that also knits female farmers together.

The couple also co-authored *Rural Renaissance: Renewing the Quest for the Good Life* (New Society Publishers, 2009), a guide to budding farmers and filled with gorgeous photos (Ivanko is a professional photographer). And they also host an Aspiring Innkeeper Working Retreat for wannabe B&B owners.

Can't make it for the night? The couple's most recent venture is their bakery, selling Latvian sourdough rye bread, Latvian piparkukas, farmstead butterhorn rolls, Alexander Cake (Kivirist's mom's award-winning recipe, a winner in the Chicago Tribune's holiday-cookie contest), and muffins online, all folding in organically grown, local, and farm-raised ingredients. Cookies in the shape of Wisconsin are iced with green and gold frosting, for the Green Bay Packers.

Rainbow Ridge Farms Bed & Breakfast
N5732 Hauser Rd., Onalaska; 608-783-8181;
rainbowridgefarms.com

Donna Murphy and Cindy Hoehne love inviting guests to interact with the animals on their thirty-five-acre farm, with a four-room B&B (each room has a private bath). "Guests are welcome to assist in milking goats; feeding calves, pigs and baby goats; and collecting chicken eggs," says Murphy and Hoehne. "We post due dates, and guests are invited to watch baby-goat deliveries and help with newborn care. Guests may also hike up the hill with the goats, a guest favorite."

Scotch Hill Farm

17310 W. Footville Brodhead Rd., Brodhead; 608-897-4288;
scotchhillfarm.com

Troy and Dela Ends named their farmstay Innisfree. Rooms are
within a century-old farmhouse on their family's certified-organic
farm that grows and sells flowers (through a subscription service),
herbs, vegetables, and farmstead soaps. Keep an eye on the website
for news of its classes and workshops. Workdays and tours (of the
vegetable and flower gardens, farm animals, and pollinator habitat)
can also be arranged.

The Speckled Hen Inn

5525 Portage Rd., Madison; 608-244-9368; speckledheninn.com

At Tom and Heather Shannon's B&B, you can bottle-feed a lamb,
interact with llamas and sheep, learn to cook with what you grow,
and collect chicken eggs. And yet there is also luxury: a Wisconsin
cheese plate, four-course farm-to-table breakfast, and—in two of the
five rooms—a clawfoot tub or jetted tub. "We grow over thirty-five
fruits and vegetables and make our own maple syrup, apple butter,
and baked goods that we sell locally [Dane County Farmers Mar-
ket]," says Heather.

Valley Springs Farm Bed & Breakfast

E4681 County Road S, Reedsburg; 608-524-2421;
valleyspringsfarmbb.com

Sitting on a 255-acre, fourth-generation dairy and beef farm, this
B&B attracts those enchanted with homesteading. Classes in pres-
ervation/canning, tours, and UTV tours of the pasture are complemented by flourishing flower and vegetable gardens. With just three
rooms, Dorothy and Don Harms create intimacy and what you would
expect on a farm (hiking trail and a front porch with a view, for
example) but not without compromising luxury (plush robes, Netflix,
and WiFi).

Farmers Markets

Baraboo Farmers Market

Courthouse Square, between 3rd and 4th Streets and Broadway and Oak Streets, Baraboo; 608-963-9879; baraboofarmersmarket.com

The square comes alive with a full range of vegetables, maple syrup, handmade soap, eggs, farm-raised trout, berries, baked goods, and meat (chicken, pork, beef, and lamb). Orange Cat Community Farm is one boutique-farm vendor (Laura Mortimore's farm raises organic vegetables) while Four Elements Herbals—nationally distributed aromatherapy products crafted from farm-raised herbs in Baraboo—is another. Vegetable plants and hanging flower baskets are also sold. When: Wednesdays and Saturdays from 7:30 a.m. to 1 p.m., early May through late October

Beaver Dam Farmers Market

Heritage Village Shops, 1645 N. Spring St., Beaver Dam

Pick up everything from farm-fresh eggs to crafts, plus of course vegetables and herbs, at this market. Earlier in the season, you will find vegetable plants to add to your own garden, and highly aromatic handcrafted bars of soap are among the vendors' wares, too. When: Wednesdays and Saturdays from 8 a.m. to noon, early May through late October

Beloit Farmers Market

300 block of State Street and 400 block of East Grand Avenue, Beloit; 608-365-0150; downtownbeloit.com

Second only to Dane County Farmers Market in size, this market offers staples (fruits and vegetables) but also unique finds like pheasant sausages, elk meat, loose teas, dog treats, and microgreens. Prepared foods include Mexican-style sorbet from Auténtica, a family-owned food truck, and Coco's Tamales. High-quality arts and crafts are in abundance, like Beloit Pottery Company's stoneware or hand-crafted leather B & T belts. You can even sit down for a henna session! When: Saturdays from 9 a.m. to noon, May through October

Boscobel Farmers Market

Depot Park, 800 Wisconsin Ave., Boscobel; 608-375-2672; boscobelwisconsin.com

Adjacent to the historic Depot Museum in Depot Park, live music is paired with the market, for a fun, but relaxing, kick start to the weekend. Arrive hungry: baked goods, corn on the cob, pork tacos, brats, fry pies, and ice cream are also sold. Canned, pickled vegetables and jellies are also frequently for sale. When: Saturdays from 8 a.m. to noon, early May through mid-October; second and fourth Saturdays from 9 a.m. to noon, November through late April (Winter Market, at Blaine Gym, 104 E. Oak St., Boscobel)

Campus Farmers Market

Union South, 1308 W. Dayton St., Madison; 608-890-3000; union .wisc.edu

Catering to students and faculty—although anyone is welcome—this market's at Union South, a student-activity center on the University of Wisconsin-Madison campus. Among the vendors are a co-op of Monroe farmers (Parrfection Produce) and a Mount Horeb apple orchard (Munchkey Apples). Farm to Table Bags curate fruits and vegetables from that week's market, along with a recipe (samples prepared by Wisconsin Union Restaurants give you a literal taste). When: Thursdays from 10 a.m. to 2 p.m.

Capitol View Farmers Market

5901 Sharpsburg Dr., Madison; 608-218-4732; capitolviewfarmersmarket.com

As this market selling vegetables, fruits, meats, and more is held in the evening, and food trucks are in abundance, folks are encouraged to pack a blanket and beverages to enjoy with the eats you procure. Live music is also part of the experience. A park is right across the street, tucked into the Grandview Commons neighborhood on the East Side. When: Wednesdays from 3 to 7 p.m., June through October

Cross Plains Farmers Market

Kalscheur's, Highway P and Highway 14 (Main Street), Cross Plains; 608-798-3520

In addition to fruits and vegetables growers, this village west of Madison features vendors selling everything from gluten-free cupcakes for a quick sugar fix to grilled meats from Bob's BBQ Emporium. This makes it easy to eat dinner at the market. Keep an eye on the market's Facebook page for news of select dates for a monthly craft market add-on. When: Wednesdays from 3:30 to 6 p.m., June through September

Cuba City Community Market

Presidential Plaza, 108 N. Main St., Cuba City; 608-744-2152; cubacity.org

At this market you will find what you would expect—farm-fresh produce, meats, arts and crafts, eggs, and baked goods—but also the highly coveted Sinsinawa Breads. Baked ten miles from Cuba City at the Dominican Sisters of Sinsinawa's home in nearby Sinsinawa, since the 1960s and 1970s, picking up a loaf is nearly a tradition for regular shoppers to the market. Another unique find are VFW-baked pies. When: third Wednesday of the month from 4 to 7 p.m., May through September

Dane County Farmers Market

Capitol Square, 2 E. Main St., Madison; 608-455-1999; dcfm.org

As much an attraction to the city of Madison as its art museums, Camp Randall football games, zoo, and botanical gardens, the Dane County Farmers Market is also the country's largest producer-only farmers market. Planning to go? Set aside at least two hours. The first year of the market was in 1972 and it is just as vibrant now as it was back then.

Bringing together between 150 and 175 vendors, such as artisan cheesemakers, organic-vegetable farmers (many are young couples in their dream career), Hmong farmers, grass-fed beef producers, growers of cut flowers and succulents, honey makers, seasoned canners, preservers, and more, pedestrian traffic weaves in one direction around Capitol Square. (It is that crowded, imagine the mall around Christmas time!). Beyond beef, you will find bison and chicken—all made without the use of pesticides or hormones. And do not be shy about arriving hungry or picking up pastries for tomorrow's breakfast: Stella's Bakery's Hot and Spicy Cheese Bread (sells out quickly!), Paleo Mama's good-for-you cake pops, Farm Pride Bakery's "Gigando" doughnuts and Pilgrim's Pantry's scones (baked in Amish and Mennonite traditions) are among your options.

Locally made—but award-winning cheese—is a huge draw, and the vendors include Willi Lehner's Bleu Mont Dairy, whose bandaged cheddar honoring his Swiss roots is aged in his Blue Mounds cave.

After a stroll or two around the market, consider visiting artisan booths on the South end of State Street, just off the square. Whether it is a cute, hand-sewn onesie for the cutest baby in your life or a piece of jewelry to remind you of your day at the market, the quality is stunning. Another must stop: Fromagination, an adorable cheese shop just like you would find in Paris, and one of the state's best selections of Wisconsin cheeses, including orphans you can munch on while you walk. Planning to spend the day in Madison? Madison's most beloved farm-to-table restaurant (L'Etoile, with chef-owner Tory Miller at the helm, and launched by Odessa Piper during the 1970s) offers a chef's-tasting, seven-course dinner folding in much of what is sold at the market. Like the cheese shop, it is right on the square. When: Saturdays from 6:15 a.m. to 1:45 p.m., between mid-April and late November

Darlington Farmers Market

Festival Grounds Park, adjacent to Main Street Bridge on South Main Street, 145 Main St., Darlington; 608-776-3067; darlingtonwi.org

There is nothing you cannot find at this market, in Lafayette County's county seat of Darlington. Although small, it still features organically grown fruits and vegetables; baked goods; arts and crafts, and little handmade pleasures like locally crafted soaps. When: Saturdays from 8 a.m. to noon, early May through late October

Deerfield Farmers Market

Deerfield Lutheran Church's parking lot, 206 S. Main St., Deerfield; 608-239-2222; deerfieldfarmmarket.com

Pretty much everything you would scoop up at the grocery store is sold at this market, from coffee beans to eggs, plus honey and fruits and vegetables grown at local farms and gardens. Plants, handmade kitchen towels, wood carvings, soaps and cut flowers, which are also for sale, add cheer to your home and garden. Vendors rotate with the seasons, so you may find an apple orchard in fall, for example. When: Saturdays from 9 a.m. to noon, mid-May through late October

Deforest Farmers Market

Between Fireman's Park and Village Hall, DeForest Street (between Durkee Street and South Stevenson Street), DeForest; 608-846-6751; vi.deforest.wi.us

Linked with live music, food carts and weekly themes, this farmers market aims to create a fun shopping experience as you pick up items like My Red Haired Auntie dog treats, Harmony Specialty Dairy Foods cheese or Pretty Killer Cookies' whimsical frosted cookies. Pantry staples like maple syrup, honey, salsa jams and jellies join farm-fresh vegetables and fruits. You can even get potted herbs and plants for your yard. When: Tuesdays from 3:30 to 6:30 p.m., early June through late October

Dodgeville Farmers Market

Dodgeville United Methodist Church parking lot, 327 N. Iowa St., Dodgeville

In addition to whatever fruits and vegetables are just picked at local farms and gardens and in season, you will find locally raised beef, local honey, eggs, handmade cards, cornmeal, and crocheted items for the home (towels and afghans). You can also pick up pies, breads, jams, and sweet eats to eat now such as sticky buns and cinnamon rolls. When: Saturdays from 8 a.m. to noon, mid-May to mid-October

Downtown Dells Farmers Market

Riverwalk Park, 105 Broadway, Wisconsin Dells; 608-291-5157; wisdells.com

While farm-fresh fruits and vegetables represent the bulk of vendors' items for sale, there are also a few surprises, like Trail Magic Coffee Roasters' coffee blends, Big Spring Family Farm's fish, soy candles, and home décor, that truly make it a one-stop shop. You can also find honey, macarons (Macarons Boutique), and cut flowers. When: Sundays from 9:30 a.m. to 1:30 p.m., late May through mid-October.

Eastside Farmers Market

McPike Park, 202 S. Ingersoll St., Madison; eastsidefarmersmarket.com

Tucked into Madison's eclectic East Side neighborhood, vendors are just as eclectic (like a bicycle-repair stand for a quick tune-up while you shop.) You can find meats, vegetables, fruits (like apples, pears, and concord grapes), and Farmer Johns' farmstead cheese to whip up farm-fresh meals, but also score flowers, plants, and soaps—plus Chrysalis Pops' organic, artisan fruit pops, and Origin Bread's loaves baked with Wisconsin-grown grains. When: Tuesdays from 4 to 7 p.m., May through late October

El Mercadito De Centro

810 W. Badger Rd., Madison; micentro.org

A spin on this market—which debuted in 2015 on Madison's South Side, as an extension of Centro Hispano of Dane County—is that it specializes in produce and items commonly woven into Latin dishes, such as tortillas. It also seeks to model the type of market, including what is sold, you would find in Latin America. Along with produce sold by local farmers and gardeners, live music, crafts, Latin American prepared foods, and kid-friendly activities are offered. When: Wednesdays from 4 to 7 p.m., mid-February through late September

Elver Park Farmers Market

Elver Park, 1250 McKenna Blvd., Madison; madwest.org/farmers-market.html

Near the Elver Park Neighborhood Center, and in Elver Park, you will find salsa and tomato sauce to make your next meal an easy task, along with maple syrup, eggs, breads, and Squashington Farm's produce. Live music is also part of the experience, plus free bicycle safety checks while you shop. When: Saturdays from 8 a.m. to noon, mid-June through mid-September.

Ferryville Farmers Market

Sugar Creek Park, Highway 35 (Great River Road), Ferryville; ferryville.com

First things first: Get a free cup of Wonderstate Coffee (until it runs out). Venders tap into hobbies popular with locals, such as fishing and hunting gear, and Amish crafters sell their wares, too. Nuts, horseradish, honey, maple syrup, jam/jellies and vegetables are also sold, plus plants, flowers, and lawn ornaments to brighten up your home and yard. When: Saturdays from 9 a.m. to 3 p.m., third week of May through late October

Fitchburg Center Farmers Market

Agora Pavilion, 5511 East Cheryl Pkwy., Fitchburg; 608-277-2592; fitchburgmarket.wordpress.com

This longstanding market—operating since 1996—unites twenty-five or so vendors each week, selling fruits, vegetables, baked goods, and flowers. Live music adds a festive vibe. Bring your gardening questions the second and fourth Thursday of the month to the Master Gardeners booth. When: Thursdays from 3 to 6 p.m., May through October (moves to Promega BTC Atrium, 5445 East Cheryl Pkwy., early November through mid-December).

Fort Atkinson Farmers Market

31 Milwaukee Ave. E, Fort Atkinson; 920-397-9070; fortfarmersmarket.com

Live music fills the air as shoppers score cheese, eggs, meat (chicken, pork, and beef), honey, maple syrup, fresh fruits, and vegetables (many of the growers are certified-organic), and pickled and jam products. There are some less-typical items sold here, too, such as jewelry, wood crafts and textiles, marshmallows, gluten-free baked goods, and lamb meat. When: Saturdays from 8 a.m. to noon, early May through late October (Winter Market is first Saturdays of the month between January and April).

Gays Mills Farmers Market

Lions Park, Main Street and West Orin Street, Gays Mills; 608-735-4017; gaysmills.org

Whether you crave jars of pickled cucumbers, rhubarb-barbecue sauce, or fresh-out-of-the-fields strawberries, this market in a village of only five hundred residents delivers. Fresh eggs are also available and live music makes the shopping experience that much more fun. When: Wednesdays from 1 to 5 p.m., mid-May through late October

Hilldale Farmers Market

Hilldale, 726 N. Midvale Blvd., behind L.L. Bean, Madison; 608-209-7130; hilldale.com

Locals know "Hilldale" for the outdoor shopping and lifestyle center, where this market is located. This also means you can pop into Anthropologie, get your laptop examined at the Apple store, or score terra-cotta plants at Wildewood (for those plants you just bought at the market). Vendors sell fresh fruits and vegetables, such as Schroeder's apples and Apple Valley Farm's heirloom tomatoes, and floral bouquets. When: Wednesdays and Saturdays between 8 a.m. and 1 p.m., May through October

Janesville Farmers Market

65 S. River St., Janesville; janesvillefarmersmarket.com

In the heart of the downtown of Rock County's largest city (and county seat), the market makes available a mix of handcrafted items (like jewelry, art, and soap) as well as the fruits and vegetables you would expect. Honey, eggs, cheese, baked goods, and meat—along with plants for your home or garden—are also sold. Depending on the week, this may mean pumpkins from Kuffer's Pumpkins or asparagus spears from Braun's Asparagus, or produce from Vietnamese

Janesville Area Convention & Visitors Bureau / Eric Panico

farmers. When: Saturdays between 8 a.m. and 1 p.m., May through October

Lake Mills Artisan & Farmers Market

Commons Park, 100 N. Main St., Lake Mills; 608-358-3412; Legendarylakemills.com

An outreach of this charming town's non-profit Main Street Program, focused on revitalizing and preserving downtown Lake Mills, the market features everything from crafts and fine art to prepared foods, plus fruits and vegetables grown by local farmers. This is one of Jefferson County's largest farmers markets. For one Saturday in mid-December, a Winter Market (in the gymnasium at Lake Mills Middle School, 318 College St.) functions as holiday shopping for many. When: Wednesdays between 2 and 6:30 p.m., early May through mid-October

Lancaster Farmers Market

Royal Bank Parking Lot, 142 Highway 61 N., Lancaster; 608-723-2820; lancasterwichamber.weebly.com/farmers-market.html

From watermelons to crafts, and ice-cream cones from Vesterman Farms' truck, this Grant County market is a slice of small-town living and all the makers who reside there. Leafy greens in spring and gourds in the fall, for example, celebrate the seasons. Annual plants like marigolds are also sold, appealing to home gardeners. When: Thursdays from 3:30 to 7 p.m., early June through late October

Lodi Valley Farmers Market

Next to Koltes Hardware, 902 N. Main St., Lodi; 608-643-8017

Vendors shift with the seasons, based upon what has just been harvested. Sweet corn, heirloom tomatoes, beans, zucchini, Swiss chard, garlic scapes, lettuce, spinach, kale, mushrooms, asparagus, rhubarb, and cucumbers are just a few of the options. Most farmers who sell here grow without the use of pesticides or chemicals. Baked goods add a nice sugar jolt to shopping, too, as does Great Harvest Bread Company's loaves. There is even a seafood vendor. When: Fridays from 2 to 6 p.m., mid-May through late October

Marshall Farmers Market

Municipal lot, Main St., Marshall; 608-422-0428; growmarshall.com

From cheese curds to canned preserves, this market shows off the local bounty, along with typical farmers-market items like fresh fruits and vegetables, syrup, honey, eggs, and baked goods. When: Sundays from 8am to noon, Memorial Day through Labor Day

Mauston Farmers Market

Juneau County Courthouse lawn, 220 E. State St.; 608-847-4142; mauston.com

With mottoes of "locally harvested" and "locally made," that is exactly what is sold at this downtown Mauston market, whether it is skin lotion or maple syrup. Other items include honey, baked goods, flowers, eggs, preserves, plants and—of course—fruits and vegetables with local roots. Two market days per week provides flexibility in shopping. When: Tuesdays from 2 to 6 p.m.; and Saturdays from 7 a.m. to noon, early May through late October

McFarland Farmers Market
Pick 'n Save parking lot, 5709 US 51, McFarland; 608-873-9443; mcfarlandchamber.com

Offbeat items you do not normally find—like cheddar bacon kettle corn, lip bombs for chapped lips, wild rice, sugar scrubs for your skin, gem-stone jewelry, and coffee grown in Costa Rica—join fresh strawberries when in season, along with eggs, honey and beeswax products, jams and preserves, and other fruits and vegetables. In winter, the market moves to McFarland Municipal Center, 5915 Milwuakee St., McFarland. When: Thursdays from 2 to 6 p.m., early May through late October; Saturdays from 9 a.m. to 2 p.m., early November through late December (Winter Market)

Greenway Station Farmers Market
Greenway Station, 1650 Deming Way, Middleton; 608-824-9111; visitmiddleton.com

Tucked into a district with sixty shops, art galleries, jewelers, and retailers in Middleton (a western 'burb of Madison), a stop at the market may check a lot of errands off your list. Carr Valley Cheese's shop, along with national brands like Chico's and Costco, plus indie faves such as Cloth & Metal Boutique, are all here. At the market, fresh vegetables, fruits, and flowers join farm-raised meats, baked goods, jam, and cheese. When: Thursdays from 8 a.m. to 1 p.m., early June through early October

Mineral Point Farmers Market
Water Tower Park, Business 151 and Madison Street, Mineral Point; 608-967-2319; mineralpointmarket.com

Another reminder this is an artsy community is a stroll through its farmers market, which debuted in 1995. A mix of fruit and vegetable farmers (such as Shooting Star Farm's certified-organic salad greens and heirloom tomatoes) join local artisans like The Denim Dog Den (recycled-denim dog and cat toys), Happy Hunter Farms's canned

goods and Magic Light Photography's artful shots of the region.
When: Saturdays from 8:30 to 11 a.m., May through October

Monona Farmers Market

Ahuska Park, 400 E. Broadway, Monona;
mononafarmersmarket.com

Each week, twenty-four vendors sell meat, cheese, eggs, fruits, honey, and vegetables. There are also some fun finds, like spices, popcorn, salsa, and Renaisance Farms's herb-infused olive oils and vinaigrettes. Have your own garden to tend? Plants are sold at the market and baked goods (grab a cup of coffee from Crescendo Mobile Coffee Bar) are a delicious start to the day, with handcrafted, goat's-milk soaps also available. When: Sundays from 8:30 a.m. to 12:30 p.m., May through late October

Monroe Farmers Market

Historic Courthouse Square, 1016 16th Ave., Monroe;
mainstreetmonroe.org

You will find the market's vendors on the north side of the square, offering everything from sustenance (like artisan grilled-cheese sandwiches, no surprise as this is home turf for world-class cheese-makers) to groceries (such as farm-fresh eggs, local meats, honey, vegetables, and pickled products). When: Wednesdays from noon to 3 p.m. and Saturdays from 8 a.m. to 1 p.m., May through October.

Monroe Street Farmers Market

Edgewood High School parking lot, 2219 Monroe St., Madison; 608-561-8290; monroestreetfarmersmarket.org

In this near–West side Madison neighborhood, just west of Camp Randall, the market brings together food-preneurs like Bayk, Wm. Chocolate, and Bloom Bake Shop, along with food from duck-meat and grass-fed beef producers, a Hmong-family farm, a heritage-apple orchard, and Farmer Johns' cheese. Grab a Mexican lunch at El Sabor de Puebla or cold brewed coffee from Let It Ride. Knitters love Observatory Hill Farm's Corriedale sheeps' wool. When: Sundays from 9 a.m. to 1 p.m., early May through late October

Mount Horeb Farmers Market

Lawn outside Evangelical Lutheran Church, 315 E. Main St.,
Mount Horeb; mthorebfarmersmarket.com

Located in downtown Mount Horeb, this market launched in 2013
with a goal to provide "a vibrant event focused on healthy food,
healthy individuals and a healthy community." The site was intention-
ally chosen because it is an easy place for locals to access on foot
or bicycle. Vendors include Drift Coffee, Bures Berry Patch, Larry
the Gourd Guy, and Jangle Soapworks—plus decadent treats from
Whoopies, Cookies and Sweets. When: Thursdays from 3 to 6:30
p.m., early May through mid-October

New Glarus Farmers Market

Hometown Pharmacy parking lot, 1101 WI-69, New Glarus; 608-
290-3905

This market's fruit and vegetable vendors move through the seasons,
from onions in spring to pumpkins come fall. Also sold at one of the
area's newest markets are ground beef, tomatoes, herbs, honey, and

jars of salsa. Meadow Ridge Alpacas and Betty's Rag Rugs are among the artisan vendors. **When:** Fridays from 3 to 6 p.m., May through October

Northside Farmers Market

Willy Street Co-Op North's parking lot in the Northside Town Center, 2901 N. Sherman Ave., Madison; 608-695-0946

Serving the capital city's Northside neighborhood, this market has enough vendors selling prepared foods that you can easily make a brunch out of it. This includes El Sabor de Puebla's tacos, tamales, and burritos; spring rolls and Stalzy's baked goods—plus Farmer Johns' Cheese, Capri Cheese, and numerous fruit and vegetable farmers. **When:** Sundays from 8:30 a.m. to 12:30 p.m., May through October; Sundays from 10 a.m. to 1 p.m., October through December (winter market)

Oregon Farmers Market

Dorn True Value Hardware Store parking lot, 131 W. Richards Rd., Oregon

Wild rice, smoked nuts, sorghum and seed-bead jewelry are among the unique items you will find at this farmers market, in addition to what you would expect: in-season, locally grown vegetables like heirloom tomatoes or squash. Farmer Johns' Cheese is another vendor, and you can also pick up baked goods, from breads to scones. **When:** Tuesdays from 2 to 6 p.m., May through October

Platteville Farmers Market

City Park, in front of Platteville Municipal Building, 75 N. Bonson St., Platteville; 608-218-4374; plattevillefarmersmarketwi.com

From pickled canned goods to dried herbs, this is a great market to help you cook better at home, along with sales of honey, maple syrup, meat (beef and pork) jams and jellies, and, of course, farm-fresh produce, herbs, and fruits. Vendors also sell their baked goods, arts, and crafts (such as soap). **When:** Saturdays from 8 a.m. to noon, May through October (Winter Market between November and April)

Portage Farmers Market
Commerce Plaza, 303 W. Wisconsin St. (corner of Highways 16 and 33), Portage; 608-742-6242; portagewi.com

Located in downtown Portage, the market's vendors rotate with the seasons, reflecting what is ripest and freshest that week, whether it is apples in September or tomatoes in August. The park's green space provides spots to linger before or after you shop or drop into Two Rivers Coffee Company's cafe across the street. When: Thursdays from noon to 5 p.m., May through October

Poynette Area Farmers Market
Pauquette Park, 106 S. Main St., Poynette; Poynette-wi.gov

Hosted in this 3.6-acre park, across from the police station, there is plenty of room for an impromptu picnic. The market sells arts and crafts, perennial and annual plants, baked goods, honey and, of course, locally grown and in-season fruits and vegetables. When: Saturdays from 8 a.m. to 11 a.m., early May through late September

Prairie Street Farmers Market
Prairie Street, Downtown Prairie Du Chien; 608-379-0801

From pork to seedlings, you can get more than fruits and vegetables here, plus baked goods to get your weekend off on the right (and perhaps a sweet?) note. Canned goods, soap and houseplants make great last-minute gifts. Check the market's Facebook page for news of which farmers will be selling that week. When: Saturdays from 8 a.m. to 1 p.m., early May through mid-October

Reedsburg Area Medical Center Farmers Market
Reedsburg Area Medical Center, 2000 N. Dewey Ave., Reedsburg; 608-524-6487; ramchealth.com

Employees of this medical center have it made, they do not even need to hop in a car to shop for fresh fruits and vegetables, as the market is right on campus. Pickled items, fresh fruits and vegetables, and baked goods are among the items sold here by a mix of vendors, farmers, and gardeners. When: Fridays from 10 a.m. to 3 p.m., May through October

Richland Area Farmers Market

Phoenix Center parking lot, 100 S. Orange St., Richland Center; 608-213-3374; richlandareafarmersmarket.org

A nice bonus for this market is it is held twice per week (Wednesdays and Saturdays), giving you no excuse to miss it. There is also a strict rule that every vendor must have grown or created what they are selling. This results in locally grown and harvested fresh fruits and vegetables, eggs, meats, maple syrup, honey, baked goods, and arts and crafts creations. When: Wednesdays from 1:30 to 5:30 p.m.

Rock County Farmers Market

5013 Wisconsin Trunk Highway 11, Janesville; 608-449-4900; rockcountyfarmersmarket.com

Sponsored by Nature's Touch Garden Center, which is the host site for the market, farmers sell cut flowers, meat, vegetables, crafts, jams, fruit, kettle corn, gourmet dog treats, and cheese. You can even get your culinary knives sharpened while you shop. Pop into the garden center after to fill out your garden at home or pick up fresh herb plants. When: Sundays from 9 a.m. to 1 p.m., early May through late October

Sauk Prairie Farmers Market

Water Street, downtown Prairie du Sac; 608-643-4168; saukprairie.com

You can pick up more than locally grown fruits and vegetables at this market. Many other items to help fill your fridge, pantry, and home are sold, like locally roasted organic coffee beans and all-natural, hand-crafted body products (soaps and essential oils). When: Saturdays from 9 a.m. to noon, May through October

Soldiers Grove Farmers Market

Driftless Brewing Company's parking lot, US 61, 102 Sunbeam Blvd. W, Soldiers Grove

Surrounded by farmers in the state's most prolific farming region, this is one of the newest farmers markets. You can pick up cut flowers, floral wreaths, and fruits and vegetables grown within a few miles. Live music is also part of the market. That it is at Driftless Brewing Company's parking lot means a pint is within easy reach after you are doing shopping. When: Fridays from 2 to 6 p.m., seasonal

South Madison Farmers Market
southmadisonfarmersmarket.com
This market has been going strong since 2002, operating five weekly locations in Madison's South Side, not leaving a single neighborhood without access. When: Sundays from 11 a.m. to 3 p.m., late April through late October, Madison Labor Temple, 1602 S. Park St., Madison; Mondays from 2 to 6 p.m., mid-June through late October, Novation Center, 2500 Rimrock Rd., Madison; Tuesdays from 2 to 6 p.m., late May through late October, Madison Labor Temple; Wednesdays from 2 to 6 p.m., late June through late October, Novation Center; and Fridays from 2 to 6 p.m., late June through late October, Villager Mall, 2300 S. Park St., Madison

Sparta Farmers Market
Mueller Square, 120 N. Water St., Sparta
While this market's vendors sell the seasonal fruits and vegetables you would expect (including mushrooms), they also sell sauerkraut, pancake flours, pickles, yarn and fibers (for knitting or crocheting), soaps and body-care products, and canned jams and jellies. Check the market's Facebook page for updates on which vendors will appear on certain weeks. When: Saturdays from 8 a.m. to noon; and Wednesdays from 3 to 6 p.m.; second Saturday in May through second Saturday in October

Spring Green Farmers Market
230 E. Monroe St. (behind the library), Spring Green; 608-575-9787; springgreen.com
If you are heading into Spring Green on a Saturday for a tour of Frank Lloyd Wright's former home (Taliesin) or an outdoor show at American Players Theatre, consider squeezing in time to shop this market. Live music accompanies the experience of buying direct

from farmers and food-artisans, whether it is vegetables, sauces, or jams. When: Saturdays from 8 a.m. to noon, mid-May through mid-October

Sun Prairie Farmers Market

Cannery Square Park at Market Street, Sun Prairie; 608-825-0762; sunprairiemarket.com

The market in this Madison 'burb sells a lot of items that could make gifts—or be eaten by you. This includes Ann in a Jam jams, Doug Jenks honey, Farmer Johns' cheese curds and—for staples—Lundeen Farms pork and chicken, and Wells Farms's beef, plus organic vegetables from Emerald Meadows Family Farm's 220-acre property. When: Saturdays from 7 a.m. to 11 a.m., early May through late October. There is also a Winter Market, early November through late April, at an indoor location.

Stoughton Community Farmers Market

Forrest Street, Stoughton; stoughtoncommunityfarmersmarket.org

Founded in 2015, this market in downtown Stoughton is near attractions like the Norwegian Heritage Center as well as Nauti Norske (Saturday lunch menu includes chilled soups like Thai Carrot or Watermelon Strawberry). Food vendors range from Elmer Meats to Pleasant Springs Fish Hatchery, plus numerous small family-owned vegetable, and fruit farms. When: Saturdays from 8:30 a.m. to noon, mid-June through late September

Verona Downtown Farmers Market

Hometown Junction Park, corner of South Main and West Railroad Streets, Verona; 608-845-5777; Chamber of Commerce building, 120 W. Verona Ave., Verona (Winter Market); veronadowntownfarmersmarket.com

Access via the Military Ridge State Trail if you wish. Vendors include The Siamese Farmer, La Ferme Dans La Vallee (duck meat and chicken eggs), and Outdoor Addiction (taxidermy and butcher shop), plus Southeast Asian condiments and CBD products. When: Wednesdays from 3 to 6:30 p.m., mid-May through late September;

Winter Market: third Saturday of the month, 11 a.m. to 2 p.m., November through March

Viroqua Farmers Market

Western Technical College parking lot, 220 S. Main St., Viroqua; viroqua-wisconsin.com

Located in the Driftless Region's largest town—where farmers growing sustainably are quite prolific—there is not something you cannot find at this market. This includes seasonal fruits and berries, meats, cheeses, duck eggs, root crops, herbs, and greens, for fifty vendors total. Artists and crafters are here, too, selling soap, photography, woodworking art, baskets, and hand-sewn items. Grab a baked good to munch on while you shop. When: Saturdays from 8 a.m. to noon, early May through late October

Waunakee Area Farmers Market

Waun-a-Bowl/Rocky Rococo parking lot, 301 S. Century Ave., Waunakee; waunakeechamber.com

This Madison 'burb's farmers market has a goal "to preserve Wisconsin's unique agricultural heritage" but also folds in artists and crafters to sell their wares. This includes jars of dilly beans, dozens of tomato varieties, bunches of carrots, and baked goods such as brownies and cookies. When: Wednesdays from 3 to 6 p.m., early June through late October

Waupun Farmers Market

111 E. Main St. (O'Connor, Wells & Vander Werff LLC parking lot; next to city hall), Waupun; cityofwaupun.org

The twenty-some vendors at this market sell their grass-fed beef, soap, pickled and preserved canned vegetables, cut flowers, seasonal fruits and vegetables, and woodworking art. When: Saturdays from 8 a.m. to noon, early June through late October

Westside Community Market

UW Health Digestive Center, 750 University Row, Madison; 608-628-8879; westsidecommunitymarket.org

The word *community* in the market's name is deeply reflective of the fifty or so vendors, who bring their meat, cheese, baked goods, pickles, jams, honey, and maple syrup each week. The goal, says the market's organizers, is to offer "curated product diversity." Cut flowers and plants for your home and garden are also available. When: Saturdays from 7 a.m. to 12:30 p.m., mid-April through early November

Folk Schools

Driftless Folk School

E13858 County Road D, La Farge; 608-620-3234; driftlessfolkschool.org

Because the Driftless Region's home to the state's highest concentration of organic farmers and people working in timeless crafts, it makes sense a school on a 120-acre farm would pass along those skills. Farm-steading and land-stewardship principles are folded into classes that teach basket coiling, shibori dyeing, butchering, cheese-making, foraging for edibles, baking bread, soapmaking, black-smithing, beekeeping, mastering hard cider, and crafting fermented beverages, and vegetables. Plans are in the works to offer camping and lodging. For now, students stay nearby.

Folklore Village

3210 County Road BB, Dodgeville; 608-924-4000; folklorevillage.org

Open since 1968, this folk school taps into a Scandinavian theme with its arts and crafts instruction (including woven paper heart baskets, building a Finnish kantele, or felting slippers), although food is where it skews worldwide with focuses on Cajun cooking or Japanese comfort foods. Nordic-inspired herbals as well as sourdough baking, and embroidered holiday ornaments are two other recent class topics. Overnight accommodations can be tacked onto registrations for a surprisingly affordable rate (around $10 for a camping site or $15 for a rustic bunk).

Shake Rag Alley Center for the Arts

18 Shake Rag St., Mineral Point; 608-987-3292; shakeragalley.org
Shake Rag Alley's two-and-a-half-acre campus gives you the best of both worlds: a six-minute walk to a brew pub but also acess to Wisconsin's countryside. Day-long and weekend workshops between March and November are taught by both local folks and those arriving from other states, and highly skilled in their craft. Topics include sewing, working with bent willow, ceramics and pottery, printmaking, writing, paper and book arts, fiber arts, blacksmithing, and coppersmithing. Lodging is in the campus's historic Coach House or on Commerce Street two blocks away.

Pick Your Owns

Appleberry Farm

8079 Mauer Rd., Cross Plains; 608-798-2780;
theappleberryfarm.com
You-pick dates for apples kick off at this family farm in late August, with brats hot off the grill and donuts just out of the oven, along with cider to wash it all down. Earlier in the summer, strawberries and raspberries can be plucked off the vines, too. Visitors are welcome to sit by the duck pond or stroll through the oak savannah (cue pretty Blue Mounds views). Hayrides are also offered. Around Thanksgiving, you can order the family's apple pies or cider.

The Berry Farmer

E1022 Hoot Owl Valley Rd., Baraboo; 608-355-1965;
theberryfarmer.com
John and Jean Pinkston grow three crops: peas, strawberries, and blueberries. All are available for you-pick, with peas and strawberries ripening between early June and early July, followed by blueberries in late July through mid-August. Keep an eye on its website and Facebook page for news of picking dates and hours—plus what

is ripe and when. The blueberry fields, for example, are open only Wednesday evenings and Sunday during the day, while strawberries and peas can be picked daily.

Blue Skies Farm
10320 N. Crocker Rd., Brooklyn; 608-455-2803 (picking info) or 608-335-7825 (produce orders); blueskiesfarm.com

I was a new Madison resident and craving farm-fresh fruit one hot August day when an online search for you-pick raspberries brought me to Paul and Louise Maki's Blue Skies, a sustainable operation since 1991, using a well-thought-out compost recipe with alfalfa meal and neighbors' llama and horse manure. To keep the farm humming year-round, the couple harvests, for example, asparagus in spring, twenty-five varieties of heirloom tomatoes (as part of its "Mediterranean line" in July, apples in August, and ginger and turmeric during autumn.

Bures Berry Patch
3760 W. Brigham Rd., Barneveld; 608-924-1404; buresberrypatch.com

Plump, bright-red strawberries can be picked from the fields in late June through early July, and with all the other items Kathy and Ed Bures grow and sell you can nearly whip up a dish with your bounty. This includes maple syrup, rhubarb, raspberries, sweet corn, squash, green beans, brown eggs, baby potatoes and local honey. The Bures have managed this farm since 2001. Pick-your-own peas and pumpkins are also available when in season. Check the website's "Now in the Patch" page for this year's ripening periods.

Carandale Fruit Farm
1046 Tipperary Rd., Oregon; carandalefarm.com

Family owned by Dale and Cindy Secher since 1969—and now by their son Cory Secher, after a career fighting wildland fires in Colorado and Wisconsin—this farm grows strawberries, aronia, pear varieties, plums, and Concord grapes. It has even engaged in "uncommon fruit trials," such as growing European black currants, which you can read all about on their website. You-pick strawberries in late June through early July are monitored through "daily updates" on the farm's website and Facebook page.

Eplegaarden
2227 Fitchburg Rd., Fitchburg; 608-845-5966; eplegaarden.com
Specializing in growing apples and pumpkins, this orchard—only ten miles from downtown Madison and owned by Vern and Betty Forest since the 1980s, with two farm managers now tending to the trees—bakes what it calls "addictive" cider donuts and presses cider weekly. These are sold at the farm store, along with honey produced by resident bees. You-pick options exist for apples, pumpkins, and raspberries when in season.

Lapacek's Orchard
N1959 Kroncke Rd., Poynette; 608-635-4780; lapaceksorchard.com
This apple orchard twenty-five miles north of Madison grows sixty or so varieties, including its most popular: Zestar! Owned by three generations of Lapaceks—including quilter Kim and her husband Jared—the apple-picking calendar on their website is kept current. Because the orchard's become a popular photography site—for both quick snaps and professional sessions—the owners rent out apple crates and charge a small (around $30 per hour) fee. Sold on site are caramel-apple coating, pressed cider, dried honeycrisp chips, and frozen apple-cider donuts. A Poynette graphics company created an apparel line, also sold at the orchard.

Mitchell Vineyard
4252 Sunny Ridge Rd., Oregon; 608-225-9210; mitchell-vineyard.com
This is a vineyard that home winemakers need to know about. Come and pick wine grapes during the harvest between early September and early October (Tuesdays, Saturdays, and Sundays) to make your own wine. They will even crush and press grapes at no cost (unless you have your own crusher de-stemmer, doubtful!), and invite you to help out for extra learning. Upon arrival, you are handed clippers

and told which rows feature the grapes (such as St. Pepin, Muscat, Delaware, Foch, Frontenac, and Marquette) you wish to pick.

The Tree Farm

8454 WI-19, Cross Plains; 608-798-2286; thetreefarm.net

It is not uncommon for fruit farms to open up their fields for you-pick days, but vegetable farms? Rarely. That is what makes The Tree Farm unique. On a random day in late August, thirty items were reported on the farm's website to be ready for picking, from basil to zucchini and including fun finds like *cucuzzi* (an Italian summer squash, also called pale green snake gourd) and *gongura* (edible leaves commonly used for Indiana cuisine). The farm is open Thursday through Sunday for picking.

Tree Farms

Silent Night Evergreens

W6717 County Road P, Endeavor; 608-587-2445; silentnightevergreens.com

Nestled in the appropriate-sounding town of Endeavor, this seven-hundred-acre tree farm that dates to the late 1970s grows—and makes available for cut-your-own near the holidays—balsam fir, fraser fir, white pine, and scotch pine. Jim and Diane Chapman's trees have even stood proud within the White House, during Presidents Bill Clinton's and George W. Bush's terms.

Summers Christmas Tree Farm

4610 Rocky Dell Rd., Middleton; 608-831-4414; summerschristmasfarm.com

Folks in Madison need not drive far for their tree each year, as this 130-acre tree farm owned and run by Bill and Judy Summers is in the western 'burb of Middleton. A gift shop sells additional decor to make your house feel like Christmas, plus logoed wear that includes a three-quarters-sleeve jersey tee depicting a red pick-up carrying a tee. Wreaths, garland, and boughs are also sold.

Bailey's Run Vineyard & Winery

N8523 Klitzke Rd., New Glarus; 608-496-1966;
baileysrunvineyard.com

Moving beyond tastings, this winery—with an on-site vineyard snug
in the Upper Mississippi River Valley AVA—hosts quirky events like
Bingo along with tasty pairings (Neapolitan-style, build-your own
pizzas on Pizza Nights each weekend) and live music (also on weekends). Tastings of the wines, crafted from mostly cold-climate grapes
like La Crescent and Edelweiss, plus flown or trucked-in more
familiar grapes like Carménère (California) and Riesling (Chile), are
lumped into five categories: sweet, dry, mixed wines, sparkling and
mixed with sparkling. Illinois grapes are in the trio of sparkling wines.

wine in Monterey, California—opened their tasting room back in 2015, pouring their wines crafted from grapes grown outside of the region (Washington's Columbia Valley, Michigan's Lake Michigan Shore and New York's Finger Lakes—plus areas within Wisconsin). The tasting room and winery's 6,000-square-foot contemporary, barn-like design is on 23 acres, with pet-friendly green space, live music (Wednesdays and Saturdays in summer), a driving range and shared-foods menu.

Baraboo Bluff Winery

E9120 Terrytown Rd., Baraboo; 608-393-0939; barabobluffwinery.com

This family-owned winery (same owners as Broken Bottle Winery) boasts the kind of vineyard views you could linger at all day—and should. Pets are welcome and so are picnic lunches. Wines crafted by winemaker Fred Quandt range from recent 2019 Finger Lakes International Wine Competition Double Gold winner (a Sangiovese with California grapes) to a popular blend called The Girlfriend (a white-wine blend featuring Finger Lakes, New York, grapes). Many wines, however, are born out of grapes in its estate vineyard, like Petite Pearl (Petite Pearl grapes).

Botham Vineyards & Winery

8180 Langberry Rd., Barneveld; 888-478-9463;
bothamvineyards.com

You will find this estate winery's bottles in liquor stores and gas
stations (because Wisconsin's gas stations carry a ton of liquor!)
around the state but a trip to the tasting room, vineyard, and winery
tells its entire story. A visit to the property's 1900 refurbished barn
can include a tasting, tour led by founder Peter Botham (who races
vintage sports cars, a theme depicted at the winery), or private
food-and-wine experience (with local cheese or chocolate). BYO
picnics are allowed on property. The tasting room debuted in 1984,
coinciding with Botham's first vintage.

Broken Bottle Winery

S2229 Timothy Ln., Wisconsin Dells; 608-432-3786;
brokenbottlewinery.net

In the spring of 2019, this winery's tasting room—with two twenty-
eight-foot tasting bars and an outdoor seating area, aimed to resem-
ble a country-western bar—debuted. The owners, who make wine
grown from grapes at their sister vineyard (Baraboo Bluff Winery)
nearby, are of the same family. Some of their six wines (with easy-to-
remember signature names like "Summer Nights," "Rowdy Red" and
"Simple Life") have placed at the Finger Lakes International Wine &
Spirits Competition. Live music on select dates.

The Cider Farm

Tasting Room: 8216 Watts Rd., Madison; 608-217-6217;
Theciderfarm.com

Making organic apple brandy and apple cider, the tasting room is on
the West Side of Madison while the orchard's on 166 acres in Mineral Point. Hour-long orchard tours showcase the very apple trees
that produce these beverages, followed by a cider tasting. Want to
make a day of it? Tack on a cheese board and bring a blanket for a
post-tour picnic. The Cider Farm produces the only apple brandy
in the country out of certified-organic, true-cider apples. For your
next pancake breakfast, pick up the Organic Wisconsin Maple Syrup
infused with apple brandy.

Drumlin Ridge Winery

6000 River Rd., Waunakee; 608-849-9463;
drumlinridgewinery.com

With wine-label graphics that are an homage to the late Wisconsin architect Frank Lloyd Wright's Prairie School style, this family-owned winery's tasting room is just ten miles north of downtown Madison. Dane County apples go into the winery's apple cider while grapes grown in its Dane County vineyard and California's San Luis Obispo County comprise the rest of the portfolio. Are you a wine geek? Classes are held regularly, along with live-music concerts. A small-plate menu that includes local artisan cheeses and Underground Collective's smoked butcher meats cures the munchies.

Hidden Cave Cidery & Old Sugar Distillery

931 E. Main St., Madison; oldsugardistillery.com

Apples grown by local orchards—including Ski-Hi, Appleberry Farm, Apple Barn, and Kickapoo Orchard—result in Hidden Cave Cidery's ciders. Walker Fanning founded the cidery in 2018 and launched offbeat flavors like Elderflower Holy Basil, Salted Caramel, and Lemongrass Lavender. Taste with a view of the operation at the sister distillery's East Side tasting room with a patio and where cocktails are made from its spirits, such as "Gringo" (*horchata* with Old Sugar Factory Honey Liquor).

J. Henry & Sons

7794 Patton Rd., Dane; jhenryandsons.com

This is one of those "one in a million" Midwest finds: a bourbon-whiskey distillery on a farm. Because Wisconsin is not Kentucky. Heirloom red corn developed by the University of Wisconsin-Madison and grown on the family's farm is woven into its bourbon, poured at the farm's tasting room. Tours and tastings (on the porch or inside the tasting room) are offered Thursdays through Sundays. Joe and Liz Henry view this project as a way to have saved the farm, especially important as it is where Joe grew up.

Rock N Wool Winery

W7817 Drake Rd., Poynette; 608-635-4339; rocknwoolwinery.com

Wines are made from grapes either grown in the six-acre estate vineyard or nearby for a mix of dry and sweet reds and whites. Tastings join pizza, cheese, crackers, and meat, served at four bars within the former barn's tasting room (open since 2013), including one with beer taps. Do not want to go home just yet? There are spots to park campers or stake a tent. Local artisans' works are sold, too. Want to see more? Book a wine-wagon ride between the vines. Each September is the Stomp Fest celebrating the grape harvest.

Wild Hills Winery

30940 Oakridge Dr., Muscoda; 608-647-6600; wildhillswinery .com

Describing itself as an orchard, farm, estate winery and vineyard, and within the Driftless Region's Ocooch Mountain Range since 1993, this farm offers tastings on the outdoor terrace overlooking the vines or inside the tasting room. For a more immersive experience, book the "Vineyard and Tasting Tour" that takes you out onto eighty-one acres and includes four wine samples. For a more chill vibe that does not require any walking, there is the "Picnic & Wine Tasting" with a ninety-minute curated tasting in the vineyard, of not only wine but also cheese, small bites, and chocolate.

Wollersheim Winery, Wollersheim Distillery & Wollersheim Bistro

7876 WI-188, Prairie Du Sac; 608-643-6515; wollersheim.com

When during the 1980s the daughter of Wollersheim Winery founders Robert and JoAnn Wollersheim—who debuted the winery during the 1970s—married a Frenchman (Philippe Coquard) with a pedigree of winemaking, the business moved in a new direction. Philippe and Julie now run the winery, with Philippe as winemaker. But it did not start there: in 1840 Hungarian count Agoston Haraszthy planted grape vines, later founding one of California's first commercial wineries, Buena Vista Winery, still in Sonoma today. Between Haraszthy's departure and the Wollersheims' arrival, the property morphed into a dairy farm.

Included in a portfolio of reds, whites, and blush wines are three Ports. Crémant (sparkling wine) and "Little Brother" Beaujolais are made by Phillippe's younger brother in Beaujolais, France. Some wines are crafted from estate-grown grapes such as St. Pepin, LaCrosse, Marechal Foch, and Millot. Examples where this is not the case include the Chardonnay (Washington custom-grown grapes) and the Prairie Fumé (Seyval Blanc grapes come from New York).

In 2015, the winery expanded into distilled spirits (apple brandy and rye whiskey). You will find Prairie Fumé (produced since 1989 and a frequent award-winner at wine competitions, including Best of Class at the 2019 Los Angeles International Wine Competition) on restaurant menus around the state. This is the wine that put Wollersheim Winery on the map and captured the wine community's attention.

A visit to this National Historic Site is akin to a visit to Sonoma or Napa. The tasting room, built in 2008, was joined in 2019 by a bistro in the former late-1800s carriage house. Wollersheim wines are paired with eats folding in local ingredients (like Willow Creek's chorizo on "The Roma" flatbread or Keller's Kornucopia tomatoes in the heirloom tart) or its own spirits (like Rye Whiskey Cheese Dip). Julie and Philippe's son, Romain, is at the helm and fresh off culinary studies in Lyon, France. Printed on the menu are suggested Wollersheim wine pairings. A bakery turns out baguettes using Lonesome Stone Mill flours, from nearby Lone Rock. Save room for dessert: you will not want to miss out on a Brandy Old Fashioned chocolate-chip cookie or salted caramel whiskey custard.

A museum within hillside caves—renovated in portions by the Hungarian count and the dairy-farm owners—further tells the family's story as it relates to wine. It opened in 2013.

Blackberry Ridge Woolen Mill

**3776 Forshaug Rd., Mount Horeb; 608-437-3762;
blackberry-ridge.com**

Open by appointment only, this family-owned mill is a utopia for knitters not just in Wisconsin but around the country, thanks to online-ordering capability. Whether it is picking up knitting patterns, the mill's own line of spun, dyed, and hand-painted yarn skeins (born out of Merino and Corriedale Midwestern sheep), or knitting needles, nearly everything a knitter could possibly need is here. The mill will even spin yarn out of wool for those who raise sheep for their fibers and fleece. Visits are possible by appointment.

NORTHEAST WISCONSIN

Towns along the lakeshore between Milwaukee and the Fox Cities, then east to Door County, embrace their Norwegian heritage. Driving northeast up I-43 you will spot towns like Belgium and Luxemburg, which may throw you for a loop.

Some farmers are selling what has been grown on the family farm or in the orchard for decades, whether that's maple syrup or Door County's red-tart cherries, while others are creating a whole new business out of the past, such as Door County Creamery's and LaClare Family Creamery's recent forays into crafting goat cheese. (By the way, you can practice goat yoga at both farms.)

This part of the state hosts other cool experiential-farm visits. You can hang out with alpacas at LondonDairy Alpaca Ranch and even spend the night near alpacas at Sabamba Alpaca Ranch. Learn about Wisconsin's history as it pertains to agriculture with a stop (and lunch in the cafe) at Farm Wisconsin Discovery Center. A more rustic approach is at Washington Island Farm Museum, a series of historic, outdoor structures on the island's north side. Fiber arts are also strong, and the farms' studios must-stops for knitters, weavers, or crocheters: Bahr Creek Llamas & Fiber Studio, Hidden Valley Farm & Woolen Mill, and the aforementioned LondonDairy Alpaca Ranch.

If you are harboring a secret desire to live the farming life, think about spending a night at Justin Duell's The G Farm in Larsen, where a tiny house awaits and he is more than eager to have you assist with chores; or at Velvet Sheep Farms in Sheboygan, run by a young couple who have literally spun their dreams into this farmstead.

Door County Creamery

10653 N. Bay Shore Dr., Sister Bay; 920-854-3388;
doorcountycreamery.com

My husband's favorite sandwich is not that crusty baguette we enjoyed in a Paris park. Nor is it a too-many-meat-layers-to-count sandwich from an Italian deli in San Francisco. While those are definite runners-up, the winner is in a Door County town of five hundred residents.

Two summers ago, we elbowed our way to the back of Door County Creamery's Sister Bay cafe, deli, boutique, and creamery, enjoying our sandwiches at Sister Bay Beach. That sandwich was so good that when it began to rain, he ignored the drizzle and kept eating. For the record, my Mozzarella (think fresh mozzarella and vine-ripened tomatoes sinking into—with the help of pesto and balsamic vinegar—ciabatta batard) was just as delicious and I, too, waited out the rain for another bite. And like many vacationers in Door County, I did not ever want to forget that day—and that sandwich—so I darted back into the boutique to snap up a T-shirt with the creamery's logo.

Because creamery co-owner Jesse Johnson—with his wife, Rachael—is also a chef, every single item sold in the goat-cheese creamery's *en par* with a fine-dining restaurant. This includes goat's-milk gelato; cheese and charcuterie plates; chèvre folding in wild ramps, truffle oil, or Door County cherries; goat's-cheese curds; hard cheeses; or the aforementioned sandwiches. Brick, feta and falltium (cave-aged on cedar planks) are the hard cheeses, born out of late-fall milk. Other items sold in the boutique are goat's-milk bars crafted by Rachael, in fourteen scents that include lilac and citrus basil.

Want to meet the "kids" responsible for all this decadent goodness? Goat-yoga classes and farm tours (includes lunch from the cafe, plus a cheese and gelato tasting, and plenty of time to, as the Johnsons say, "hang" with the goats) allow cheese lovers to engage more deeply with the operation. Their farm is also in Sister Bay, just a few miles up the road from the creamery. To view Jesse when he is making the cheese, just walk up to the glass wall in the back of the cafe.

For now, the Johnsons want to keep their boutique business just that: small. This way they can travel with their kids to warm spots during the Wisconsin winters and remain engaged with their customer base.

Kelley Country Creamery

W5215 County Road B, Fond du Lac; 920-923-1715; kelleycountrycreamery.com

This sixth-generation, 220-acre farmstead scoops up ice cream, shakes, malts, and sundaes culled from their Holstein cows' milk. Choose from 22 flavors—from a roster of 267 (that is not a typo!)—with quirky, farm-appropriate names like Barnyard Bash and Cow Jumped Over the Moon, plus savory options like Jalapeño and Maple Bacon. Relax in a rocking chair on the sun porch or view ice cream churning inside.

LaClare Family Creamery

W2994 County Road HH, Malone; 920-670-0051; laclarefamilycreamery.com

Goat yoga. A restaurant where every item features cheese. A cheese shop. Goat's-milk ice cream. Are you planning a visit to LaClare Family Creamery yet?

During the 1970s Clara and Larry Hedrich began selling their goats' milk to local farmers. Then, when one of their five children (Katie) accepted an award on behalf of the family at a banquet in 2009, where she mingled with cheesemakers, she thought *Hey, we can totally do that.*

And she did. After apprenticeships with Saxon Creamery, Willow Creek and Cedar Grove (all are in Wisconsin), as well as a stint in Holland, LaClare Family Creamery's cheeses debuted, followed by the current visitor facility in 2012. Only a year prior, Evalon with Fenugreek won "Best of Class" at the US Championship Cheese Contest. Katie was only twenty-five years old. The awards kept rolling in. In 2014, Evalon earned a spot in the top-sixteen cheeses at the World Championship Cheese Contest. More recently, in 2018,

its Cave Aged Chandoka placed in the top-twenty at the World Championship Cheese Contest. In addition to chèvre and hard cheeses (including the Gouda-style Evalon, named for Larry's grandmother; and Chandoka, the original name of Larry's grandparents' farm), the creamery produces bottled goat's milk.

Katie still serves as head cheesemaker and sales and marketing manager, and now owns a butcher shop called Salchert Meats with her husband Jeff.

On a visit, buy cheese in the shop (which also retails Saxon Creamery's cheese, plus new LaClare goat's-milk lotion), sit down to lunch or dinner in the cafe (menu items include a Cubano sandwich featuring Saxon's Snowfields and Big Ed's Gouda cheeses; or a Wisco Club salad topped with LaClare's cheese curds), or take a self-guided tour of the cheese-making process thanks to a wall of windows. Between Memorial Day and Labor Day, the Kid Zone—get it, "kid" for goats?—is open for playtime with the smallest goats. "Picnic With the Kids" includes a boxed lunch and opportunities to pet goats while you picnic. And then, of course, there is goat yoga, where goats frolic among participants in a yoga class.

Brown County Fair

Brown County Fairgrounds, 1500 Ft. Howard Ave., De Pere;
920-336-7292; browncountyfair.com

From judging of animal breeds to homemade beer and wine, this fair
celebrates agriculture in this part of the state, which includes the
county seat of Green Bay. Carnival rides, a demolition derby, truck,
and tractor pull, a rodeo, and family-friendly entertainment (on a
dedicated kid's stage) are also on the schedule. You can also catch
the Oscar Mayer Wienermobile here, and all the deep-fried fair
foods you would expect. When: mid- to late-August

Calumet County Fair

Calumet County Fairgrounds, 900 Francis St., Chilton;
calumetcountyfair.com

Adopting a different theme each year—in 2019 it was the quirky
"Unfairgoatable"—there is live entertainment all four days, as well
as carnival rides and animal judging. Pop-up events at this fair that
dates back to 1856 include a car show and the hilarious "catch a
pig" contest on Labor Day, a tractor pull, Skid Steer Olympics, Ag
Olympics and a horse show. Each year the Fairest of the Fair and Jr.
Fairest of the Fair are crowned in June, and ready to serve as fair
ambassadors come August. When: Labor Day weekend

Columbia County Fair

Veterans Memorial Field, 300 Superior St., Portage;
columbiacofair.com

One of the few—if not the only—fairs in Wisconsin that is free to
enter (parking is also free, if you can believe that), activities consist
of viewing antiques and flowers displays along with judging of animals (beef, dairy, fowl, sheep, pig, goat, and rabbit). Live music shows
at the grandstand carry a fee. Crowd-favorite foods are 4-H's BBQ
and malts and Fireman's sweet corn. Carnival rides (Ferris wheel and
Tilt-A-Whirl) are also at the fairgrounds. When: last full week of July

Door County Fair

John Miles County Park, 812 N. 14th Ave., Sturgeon Bay; 920-746-7126; doorcountyfair.com

Daily admission includes access to the grandstand, fairgrounds, rides, and all events. Parking is free. Scheduled events include a demolition derby, motorcycle races, wood carvings, pig-and-duck races, and live music. Daily you will find dairy-cattle exhibits, as well as that of swine, horses, sheep, goats, poultry, and rabbits—along with domestic dogs and small animals. Food stands operated by local groups like the Lions Club, 4-H, and Farm Bureau sell brats, walking tacos, root-beer floats, breakfast (Friday and Saturday) and more. When: late July through early August

Florence County Fair

Florence County Fair Park, 5505 County Rd. N, Florence; 906-282-9153; florencecountyfair.com

Live music, a garden-tractor pull, draft-horse pull, the Mud Bog, and classic-car show are among the fair's events. Operating since 1903, the fair has added a few items in recent years, including a polka mass and Secret Life of My Pets tea party. Like all county fairs, livestock judging as well as judging of non-livestock items (for example, crafts) are also woven into the schedule, as are carnival rides and fair foods like a fish-fry fundraiser on Friday. Held: late August

Fond Du Lac County Fair
Fond du Lac County Fairgrounds, 520 Fond du Lac Ave., Fond du Lac; 920-929-3169; fonddulaccountyfair.com
The county fair on this southern tip of Lake Winnebago has been going strong since 1852. Today's events include a fourth-generation Mexican circus, the kid-friendly Barnyard Adventure Show, live music, and a fireworks show Saturday night. Animal viewing is the "cow palace" and horse barn. Local non-profit groups and mobile food vendors alike sell food. When: mid-July

Forest County Fair
Forest County Fairgrounds, corner of Glen and Railroad, Crandon; 715-478-3450
This century-old fair has exhibited animals (horses, rabbits, pigs, cows, and chickens, for instance), arts and crafts (such as hand-knit items), and culinary items (like homemade pies and farm-fresh eggs). Hot-beef sandwiches are part of a Sunday fundraiser. A horse show in the ring is a highlight. When: early September

Green Lake County Fair
570 South St., Green Lake; 920-294-4033; greenlake.extension.wisc.edu/fair
With a tagline of "best little fair around," this is the place where swine are shown, archery is judged, dogs demonstrate their agility, and there is a demolition derby. There is also something you likely will not find at another county fair in the state: agri-puppet shows. Other fair events are the Green Lake County K-9 demonstration; live entertainment that includes music, comedy, or illusion (magic); and a truck-tractor pull. When: early August

Kewaunee County Fair

625 3rd St., Luxemburg; kewauneecountyfair.com

In 2017, this fair celebrated its centennial, building on a century of farming families in this county on the shores of Lake Michigan just south of Door County. Carnival rides include The High Roller and The Zipper and the grandstand hosts hard-rock shows, while activities with less adrenaline draw just as many crowds: a tractor pull, the "Crown & Sash Dash" (participants recycle formal wear in a race), and Junior Amateur Talent Contest. Judging of everything from flowers to livestock is also at the fair. Each year a Fairest of the Fair ambassador is crowned. When: early July

Langlade County Fair

1633 Neva Rd., Antigo; 715-216-2356; langladecountyfair.net

From offering a show (live music, demo derby, stock car, or tractor pull) to showcasing 4-H exhibits, this county of only around twenty thousand residents organizes a sizeable fair. Family-friendly shows are also part of the fair's schedule, as are carnival rides and eats that include corn dogs, funnel cakes, cheese curds, and steak sandwiches. When: late July through early August

Manitowoc County Fair

4921 Expo Dr., Manitowoc; manitowoccounty.com/fair

Each year's fair theme includes colors (for example, "Stars, Stripes and Fair Delights," with red, white, and blue hues). Many events are specific to Wisconsin, such as a home-brew competition, a cheese-carver demonstration, polka music, and cream-puff-eating competition. Judging of swine and other animals, as well as live music, bull riding, demo derby and carnival rides, are also on the schedule at this fair that has been happening for more than a century and a half. Fireworks light up the sky. When: late August

Marinette County Fair

Marinette County Fairgrounds, Fairgrounds Road, Wausaukee; 715-856-5021; marinettecountyfair.com

Hugging the west side of Green Bay, this county's fair celebrates what is made state-wide and closer to home, such as cheese curds, and also animals open for viewing inside exhibit buildings. Carnival rides include a Ferris wheel and Tilt-A-Whirl. Entertainment varies each year but typically means live bands and a hypnotist/magician, in addition to kid-friendly activities. When: late August

Marquette County Fair
Marquette County Fairgrounds, 757 Main St., Westfield; 608-296-5200
Fair organizers call this fair "a family-friendly and fun atmosphere with a small-town flavor." This includes Farm Bureau chicken dinners or Capt. Jack's BBQ washed down with old-fashioned sodas or 4-H milkshakes, or viewing a pie-tasting contest. A diverse group of animals (large and small) and arts and crafts submissions are exhibited for judging. Tractor pulls, bull riding, and live-music events fill the schedule. Not only is there a carnival, but an escape room, too. When: early July

Oconto County Fair
Oconto County Fairgrounds/Zippel Park, East Park Street, Gillett; ocontocountyfair.org
Judging of cats and dogs (alongside farm animals, both small and large), live music at the grandstand, a flea market, draft-horse, and tractor pulls, and contests in who can make the best ice-cream sundae and grilled-cheese sandwiches are among the fair's events. Farm tours are offered daily at 12:30 p.m. The fair also crowns Senior and Junior Fairest of the Fair each year. When: mid- to late-August

Outagamie County Fair
637 N. Main St., Seymour; 920-833-2941; outagamiecountyfair .com
With Appleton (a sizeable city) as the county seat, this fair—its first year was in 1885 and Willie Nelson has even performed here—has the advantage of being close to a metro area. From carnival rides to live music (acts in the Nicolet Tent are free) and truck and tractor pulls, the fair's schedule is joined by foods you would expect to see and taste at a fair as well as animal judging. When: late July

Portage County Fair

186 Forest St. West, Rosholt; 715-572-4758; roshaltfair.com
For more than nine decades, this fair has celebrated with odes to
farm life like a tractor pull, horseshoes, and animal exhibits, but
also an escape room, carnival rides, magic and hypnosis show, and
wrestling entertainment. When: Labor Day weekend

Shawano County Fair

Shawano County Fairgrounds, 990 E. Green Bay St., Shawano;
715-526-7069; shawanospeedway.net
Horses—along with animals in the dairy, livestock, and poultry-rabbit
barns—are judged at this fair, and non-animal exhibits (crafts, etc.),
too. Live entertainment at night spans hard rock and hypnotism.
There are also kid-friendly carnival rides and kiddie tractor pulls,
with Thursday fireworks to kick things off. Go-Kart races are at the
racetrack. When: Labor Day weekend

Sheboygan County Fair

229 Fairview Dr., Plymouth; 920-893-5751; shebcofair.com
Entertainment over the five-day fair includes pig and duck races,
family-friendly activities in the ag-venture tent, animals and crafts
judging, and performances by dock dogs (canines diving off docks).
Food options such as cotton candy, egg rolls, walking tacos, and
steak sandwiches on Sheboygan hard rolls (it is a Sheboygan thing!)
are tasty traditions. Each year's Fairest of the Fair serves as the fair's
ambassador. When: Labor Day weekend

Waushara County Fair

Waushara County Fairgrounds, 513 S. Fair St., Wautoma;
920-647-0694; wausharacofair.com
Folding in a livestock parade; horse racing; demolition derby; and
exhibits of animals, fruits, flowers, and crafts, there is something for
everyone at this fair. There are also carnival rides, a magician enter-
tainer, a truck and tractor pull, and live music at the grandstand and
Midway Music Hall. When: late August

Winnebago County Fair

Sunnyview Exposition Center, 500 E. County Rd. Y, Oshkosh; 920-685-3013; winnebagocountyfaironline.com

A rodeo, truck and tractor pull and demolition derby each year join live bands, carnival rides, kid-friendly entertainers, plus exhibitors of farm animals, arts, and crafts. Unique from other county fairs is that you can camp out all week in a tent, RV, or trailer for about $75 to $150. When: early August

Dairy Centers

Lamers Dairy

N410 Speel School Rd., Appleton; 920-830-0980; lamersdairyinc.com

The store at this family-owned dairy founded in 1913 evokes nostalgia. Since 1998 it has been tucked into a red barn at this location. Just past the blue silo, you can snap up the dairy's rBGH-free milk (sourced from family farms within twenty miles) in returnable glass bottles, plus Wisconsin cheese and sausages from various producers. Come fall, there is pumpkin-spice egg nog. For a taste of the Lamers' dairy success, order a soft-serve ice-cream cone. Check out the thirty-foot-tall mural the family commissioned from artist Sheri Jo Posselt on their one-hundredth anniversary.

Farm Museums

Farm Wisconsin Discovery Center

Farm Wisconsin, 7001 Gass Lake Rd., Manitowoc; 920-726-6000; farmwisconsin.org

In many ways, this ten-thousand-square-foot educational center—in partnership with Lakeshore Technical College—devoted to Wisconsin agriculture is like a mini state fair. Up to two calves daily are born in the Land O'Lakes birthing barn. Off-site farm tours via coach-bus transportation are included with the cost of admission. Back at the center, exhibits showcase how Wisconsin agricultural producers make food, fiber, and fuel while tapping into safe, humane, and sustainable methods. If you are wondering about the depth of what is grown in Wisconsin, the "Wisconsin's Diverse Agriculture – From Alfalfa to Zucchini" has you covered, while "Common Ground – Balancing Farming and the Environment" shares all the ways the state's farmers preserve and protect natural resources even while repurposing the soil for food production.

Arrive hungry because the adjacent indoor cafe (plus an outdoor patio) is a showcase for all that can be grown in Wisconsin for edible consumption. Open for breakfast and lunch (from 8:30 a.m. to 2:30 p.m. daily), the menu includes Barnyard Bowl (organic Milos Farm's eggs tossed with potatoes, vegetables, and Sargento cheddar and served with toast or bread and Door County preserves) and grass-fed Angus burgers from Bailey's Harbor in Door County (meat-free eaters can enjoy the walnut burgers, another Wisconsin specialty). Henning's cheese curds are even dipped in sweet batter and lightly fried, a mix between blueberry pancakes and, well, cheese curds. Sixteen flavors of ice cream come from Cedar Crest Ice Cream, a local creamery.

Save the shopping for last. Packed into the gift shop are a mix of culinary products and Wisconsin-pride items—all with local roots. Blocks of artisan cheese from award-winning creameries as well as jars of maple syrup, bottles of wine and craft beer, and locally made sodas are practically enough to pull together a picnic.

Historic Farm Museum
1701 12th St., Two Rivers; 920-553-4001; tworivers-history.org
Managed by the Two Rivers Historical Society in this small Manitowoc County town along Lake Michigan, this museum helps tell the story of how far farming has come. There is a special focus on dairy farming, which continues to be prominent in the region but was even more so years ago. View wood carvings by former dairy farmer Dick Hazaert, gaze into rooms at a replica 1800s farmhouse, and ogle at antique tractors. Admission is free and the museum is only open in June, July, and August.

Washington Island Farm Museum

1675 Jackson Harbor Rd., Washington Island
Self-guided tours of nine outdoor structures—utilized between the 1870s and 1940, until lovingly restored by volunteers—on twenty acres allow you to visualize life as a farmer around the time these buildings popped up. Stocked with vintage farming and blacksmithing tools, farm animals and a period-furnished log cabin complete the vibe. On Wednesdays, learn to weave rag rugs and take a wagon ride. A farmers market is on Thursdays. Look for family-friendly crafts on certain days. Columbus Day weekend is an annual cider-pressing party. Visitors can picnic on site.

Farms and Farm Stands

Avrom Farm

W0908 Scott Hill Rd., Ripon; 920-204-7594; avromfarm.com
After graduating from Warren Wilson College in 2017, Hayden Holbert rebooted the farm his grandfather—a Ripon College art professor whose steel sculptures adorn the property—bought in 1950, raising vegetables, herbs, pork, and chicken through regenerative farming practices. Holbert grew up in Chicago, where he operates two pick-up sites. You can also pick up at the farm. Farm-to-table dinners and a family-friendly music and food festival are during summer. Holbert curates from other Wisconsin farms and food artisans—such as Dalla Terra Pasta dried pasta, Clover Farms beef, and Sassy Cow Creamery milk—for one-click online orders.

Bahr Creek Llamas & Fiber Studio

N1021 S. Sauk Trail Rd., Cedar Grove; 920-668-6417; bahrcreek.com
If you are a knitter, felter, spinner, weaver, or crocheter, this fifth-generation llama farm's store is your jam. An entire wall of cubby-holes is filled with yarn skeins from brands like Malabrigo (merino and alpaca from Peru) to the farm's own fibers. You can also find hooks, bamboo needles, spinning wheels, and looms. The owners invite you to linger, and perhaps work on your project, with seating

around a vintage wooden table. Are you a beginner? Sign up for knitting classes. Pre-knit garments like scarves and hats are also sold.

Barnard Farms
5807 State Highway 42, Sturgeon Bay;
pickyourowndoorcountycherries.com
Owners Crystal and Jim Barnard have grown and harvested fruit—strawberries, apricots, peaches, apples, plums, and cherries—since the 1970s but it wasn't until 2011 that they launched their own stand in Carlsville, at the southern tip of Door County's peninsula. They are proud to own the county's oldest farm market, once operated by the Krefts. Around the third week of July is when cherries ripen and you-pick options debut in the fields, which are farmed using integrated pest management (pesticide-free).

Cherry De-Lite/Country Ovens
229 E. Main St., Forestville; 920-856-6767; countryovens.com
One of my favorite items in my pantry is Cherry De-Lite sprinkles, sold at Mike and Kathy Johnson's factory-retail store. I spoon these Montmorency tart-cherry sprinkles on top of ice cream and yogurt to extend the tart-cherry season past summer. A close second is the brand's cherry syrup, which easily sweetens up pancakes or waffles. You can also pick up dried tart cherries, cherry juice, cherry-pie filling, cherry vinaigrette, cherry BBQ sauce, cherry chutney, cherry-jalapeño salsa, cherry mustard, and chocolate-covered cherry snacks. The only reason you wouldn't go here? You somehow do not like cherries.

Cuff Farms
N2299 Ledge Hill Rd., Hortonville; 920-779-4788; cuffarms.com
With a cute tractor label on all its canned jams, honey, maple syrup and pickled products, this sixth-generation strawberry and pumpkin farm (dating back to 1849) hosts an annual farm-to-table dinner as well as a petting zoo of small animals that kids enjoy. During summer, locals flock here for pick-your-own strawberries. Come autumn, the pumpkin patch is open for picking and pizzas as well as BBQ are baked in a wood-fired oven.

Door County Underground

920-421-3580; doorcountyunderground.com

As its name suggests, to experience one of Door County Underground's pop-up dinners in the farm fields and forests of Door County requires a little bit of sleuthing. This includes staying in the know by frequently consulting the calendar on its website or following on Instagram. Most dinners, however, are hosted at Hidden Acres Farm in Sister Bay although some are at other farms.

This is a totally mobile business, co-owned by Jamie Mead and Matt Chambas who work with Hidden Acres Farm's owners on curating a memorable feast. Raised in Madison, Chambas cooked with two of that city's most renowned chefs: James Beard Award–winning chef Tory Miller (L'Etoile) and Shinji Muramoto (Muramoto), plus spent time in the kitchen at two fine-dining spots in Door County (Wickman House in Ellison Bay, and Trixie's (in Ephraim). Mead is a Door County native who cut her chops in the family business: The Shoreline Restaurant in Sturgeon Bay. Chambas is obsessed with wildcrafting and foraging, which comes through in every meal, and Mead is an avid forager, too. They are continually in pursuit of delicious ingredients with not only local roots but medicinal qualities.

Each dinner—which is anywhere from five to thirteen courses, plus beverage pairings—is hosted in an intimate setting and entirely about cooking and plating organic, seasonal, and locally foraged ingredients. "We offer dining experiences that draw you back to the root of the food," say Matt and Jamie. What does this mean? By only cooking with foods in season, these are the freshest options and a love letter to Door County's fertile soil. Many ingredients are grown at Mead's and Chamba's organic garden, which they dub "Red Cabin Farm." One of their favorite ingredients are aronia berries from Hidden Acres Farm, celebrated for their "acai-like superpowers," they say.

Partnering with guest chefs invites new influences on the dinners and cooking classes are offered on occasion. Soon they hope to sell their own line of sauces, conserved mushrooms, and pickled vegetables.

Englewood Grass Farm

W1414 County Road Z, Fall River; 920-484-8457;
Englewoodgrassfarm.com

At this grass-fed beef farm is an event most farms do not host—nor
do they offer activities with this much adrenaline. Each spring the
farm hosts a national bike race (one of ten USA Cycling events
around the country, referred to as Englewood Open) that attracts
elite and pro mountain-bikers from outside the area. Many of these
cyclists are preparing for upcoming Olympics races or World Cham-
pionships. A Wisconsin Off Road Series race (part of the country's
largest state mountain-biking series) is also at the farm.

Floppy Ear Farm

16711 Hilltop Rd., Reedsville; 920-775-9364; floppyearfarm.com

Christine Kocourek's and Keith Schroeder's body-care products
from their pasture-raised goats' milk— soap bars, bath bombs, "bug
bars," liquid soaps, and lotions—have won awards from the American
Dairy Goat Association. Take a self-guided barn tour and visit the
forty-six-acre farm's shop (check the website for that season's days).
"People can see and pet the yaks (Molly and Lily), a selection of
goats, our mule, our mammoth donkey," says Kocourek. On the first
weekends in May and December, an open house brings in additional
vendors plus guided barn tours.

Fragrant Isle Lavender Farm, Shop & Le Cafe

1350 Airport Rd., Washington Island; 920-847-2950;
fragrantisle.com

You may expect a couple who used to live in Miami and Chicago to
buy a vacation property in an exotic, metropolitan locale. Martine
and Edgar Anderson actually had their hearts set on rural Washington Island, a community of around seven hundred people where few
besides them live here year-round. A friend's cabin served as their
introduction.

After an early retirement—she worked in merchandising for
luxury retailers like Saks Fifth Avenue and Neiman Marcus, while
he flew around the world managing McDonald's construction

projects—they moved into their Door County home. Only they were not ready to stop working.

Martine wanted to create a living memory of her growing up in the South of France. The Midwest's largest lavender farm—with twenty thousand lavender plants, spanning a dozen varieties on 21.7 acres since its 2012 opening—blends their passions beautifully. "They wanted something that was good for the environment, good for the Washington Island economy, and something that would be useful to someone in their everyday life," says marketing director Julie Imig. Walking into the boutique, you will spot vignettes of retail products—from books about French culture to cute ballerina dresses—obviously curated by Martine, guided by lavenders' calming scent. The shop also sells the farm's own line of bath and body-care products as well as sachets and a new CBD line. In the cafe are sold macaron cookies, sandwiches, ice cream, chocolates, and glasses of wine, beer, lavender lemonade, and a twist on the ever-popular Frosé (a lavender *Rosé* slushie).

It is rare for someone to spend less than an hour here, what with the English-style lavender gardens, white wrought-iron structures, and beckoning Adirondack chairs.

"Every year, they are doing things to enhance the lavender experience," says Imig, gesturing to the new structures and chairs that had recently arrived. The Andersons even tested lavender in their home garden before plotting out the farm. In July and August are you-pick lavender days in the fields and if you are lucky, you may arrive on a day when distilling is being done. This therapeutic-grade lavender is then used to craft products like lotions, balms, and sprays.

A second retail shop has been in Fish Creek, at 9341 Spring Hill Rd. since 2017, providing access to lavender products to those on the peninsula.

Hidden Acres Farm
11128 Beach Rd., Sister Bay; hiddenacresdoorcounty.com
Tom and Carolyn Rehberger's vegetable farm is a frequent host for Door County Underground's pop-up, multi-course dinners celebrating locally grown, in-season, and wild-foraged foods. The Rehbergers also supply local restaurants that include Base Camp Coffee Bar (Sister Bay), Wickman House (Ellison Bay), both Wild Tomato locations, and Roots Inn and Kitchen (Sister Bay). A farm stand ensures organic produce, floral bouquets,and aronia berries (available for you-pick in September) are never out of reach. Twenty-six community-garden plots teach others the love of farming.

Hidden Valley Farm & Woolen Mill
14804 Newton Rd., Valders; 920-758-2803;
hiddenvalleyfarmwoolenmill.com
"Any given time of year we have four hundred to six hundred sheep, depending on what is transpiring," says Carol Wagner, co-owner with her husband, a former Peace Corps volunteer who grew up on the land. They moved here from California to start the business during the 1980s. "We do everything: we raise the sheep, process the wool, direct market the meat, and sell the fibers." Indeed, the shop sells yarn skeins, raw fibers, and even antique spinning wheels. A holiday sale the weekend before Thanksgiving and barn sale the first weekend in May attracts visitors, too.

Inthewoods Sugar Bush

1040 S. Union Rd., Manitowoc; 920-242-9050; inthewoodssugarbush.com

In 2020 the Wagners unveiled a brand-new sugar shack on their seventeen-acre farm, where 1,300 maple trees have been tapped for their sap (in all grades). Not only can you purchase the farm's maple syrup, but also maple-syrup candles, maple sugar, and maple cream. Visitors are welcome to stop by—just call first—during the sugaring season (typically late February to mid-April), to witness firsthand the process of making maple syrup.

Island Lavender

1309 Range Line Rd., Washington Island; 920-737-1531; islandlavender.com

Sprawling retail shops inside the former dairy serve as a highly aromatic education about lavender's adaptations, such as sipping tea, blending into syrup, packaging into soap bars, bundling into sachets or crafting into truffles. It is free to roam the grounds, where Norwegian flags wave above fields of lavender (first planted in 2013) and a small historic museum is in a replica Stave church just like you would find in Norway. A second store is in Ephraim, at 10432 Highway 42 (Water St.), and another in Beaufort, South Carolina.

Kelley's Farmstead

N4149 Kelley Rd., Fond du Lac; 920-979-0955; kelleysfarmstead.com

From the Kelleys who own and operate Kelley Country Creamery (also in Fond du Lac), this farmstead—which runs on sustainable-agricultural philosophies, harvesting corn, wheat, alfalfa, and sunflowers—comes alive each fall with corn mazes, grain-cart rides, a beautiful field of sunflowers, and games of pumpkin checkers and tic-tac-toe (weekends in September and October). Pumpkins, gourds, and ornamental ears of corn are sold, too. The farm season kicks off in April (weekends only) with animals at Spring Baby Farm Animal Barn.

Koepsel's Farm Market

9669 Highway 57, Baileys Harbor; 920-854-2433; koepsels.com

Half of this indoor farm stand—dating back to 1958 and now owned by Kevin and Karrie Oram: the third-generation owner and her great-great-great grandfather were among Door County's first settlers in 1835—sells antique furnishings while the other features anything you can imagine pickled, canned, or preserved. Items with local accents include Door County cherry-amaretto pie filling and a build-your-own six-pack of Wisconsin craft beers. Just-picked apples and pumpkins are sold here, too. Love bottled sodas in wacky flavors? You have found your place. Baked goods, Wisconsin cheese and bulk teas (including a calming lavender) round out the offerings.

Londondairy Alpaca Ranch

6827 State Highway 147, Two Rivers; 920-793-4165; londondairyalpacas.com

"They make great lawnmowers," quips Laura Prellwitz on a guided tour of LondonDairy Alpacas, on a former dairy farm dating back to 1883. "When you see them grazing, they're just sort of gnawing off the grass. They're very environmentally friendly that way."

Indeed, Destinati and Justify are going to town on the grass a few feet away.

Retired schoolteacher Kevin Stoer has always had a thing for alpacas. In 1995 he took over the property—as the farm's fourth-generation owner—and, with eight alpacas, began sharing that passion with visitors. He now raises around fifty alpacas for their fleece on eighteen acres.

Guided hour-long tours (between May and October) are filled with factoids about the breed (such as the fact that they are never wild and always owned by someone) but also fun activities like feeding the furry creatures and sorting their fibers by weight. Stoer even organizes eleven-day tours each winter to Peru for a tight-knit group of fifteen people, to experience where most of the world's alpacas live. "The Andes Mountains, that's their idea of a beach vacation," says Prellwitz, due to the high elevation.

Any lover of soft, hypoallergenic fibers will want to pop into the market—within what was once a bottling plant for milk, now playing

Andean flute music—before leaving because that's where scarves, shawls, handbags, hats, blankets, and socks are sold." Skeins of alpaca yarn in differing weights appeal to knitters wishing to craft a memory from their time on the farm. Argentinian wines are also sold here, exclusive to the area, and trinkets brought home from Peru each winter.

During the last weekend in September the farm hosts National Alpaca Farm Days, with local 4-H members on hand to answer questions about alpacas and facilitate "meet and greets" with the furry animals. Other annual events include Fiber Farm Frolic in October and Holiday Open House on select dates in November and December. During the warmer months, and into fall, activities such as yoga, sipping wine, painting, stenciling, and knitting are—pardon the pun— woven into an intimate event with the farm's alpacas.

Meuer Farm
N2564 US Highway 151, Chilton; 920-418-2676; meuerfarm.com
Whole-grain flours (out of ancient grains like einkorn, spelt, and emmer) are milled by David and Leslie Meuer on their 150-acre farm, available for pick up on site. Other items to help fill your pantry include maple syrup, honey, and meat cuts—plus fresh vegetables and fruits farmed using sustainable crop methods, like strawberries, sugar-snap peas, and sweet corn. Each fall a corn maze debuts, along with pumpkins and squash for sale. The farm also hosts educational classes in beekeeping, making maple syrup, planting strawberries with success, and growing/milling grains.

Seaquist Orchards Farm Market
11482 WI-42, Sister Bay; 920-854-4199; seaquistorchards.com
A stop at this bustling, fifth-generation-owned market along Highway 42 is practically a rite of passage for Door County visitors. It is especially busy in late July once cherries ripen and closes for the season in late October. The Seaquist family owns and manages a one-thousand-acre orchard bearing Montmorency tart cherries (around six million pounds of cherries each harvest), plus fifty acres of apple and sweet-cherry trees. Also sold in the market are baked goods (cherry pies and apple-cider donuts), plus canned products (from cherry-pie filling to cocktail cherries).

Sunny Hill Farm

1922 Oak Rd., Green Bay; 920-434-9009; sunnyhillfarm.net

Pick-your-own tomatoes is what draws people to this farm (in business since 1979) in late July through September, lured by the prospect of paying just around $20 for a five-pound pail that can easily be turned into sauce. Carrots, sugar snap peas, strawberries, beets, kohlrabi, peppers, and jalapeño are also open for picking. The farm operates vegetable stands at area locations, including local Ace Hardware and Shell stations (check the website for up-to-date information).

Sweet Mountain Farm

1402 Mountain Rd., Washington Island; 920-847-2337; sweetmountainfarm.com

The owners of this Washington Island honey and maple-syrup farm are so passionate about tending their Russian bees (a cold-hardy option for brutal Wisconsin winters) they sell supplies (like custom hives) to would-be beekeepers. A farm store retails honey, beeswax candles, Queen Bee soap bars, "icy peppermint" beeswax lip balm, and maple syrup. You can also sponsor a beehive to keep the farm humming. Find their products in the off-season at Susie's Sweets on Main Street, also on the island, and their honey woven into Door Peninsula Winery's mead wine.

The Little Farmer LLC

N9438 Highway 151, Malone; 920-921-4784; mytlf.com

Between shopping for artist-made crafts in the Craft Barn and lingering over coffee in the coffeehouse-style farmhouse, you can easily spend two hours at this apple orchard. During apple season, its Applehouse sells apples, apple butter, caramel apples (plus caramel sauce to make your own) and bakery items like pies and muffin. Keep an eye on the website for you-pick dates for apples and info about farm tours. A playground, corn maze, goat feedings, the Peachick Truck (chickens' home), sheep and hayrides keep restless kids happy—there is even a dedicated diaper-changing pavilion to prove how family-friendly it is.

Waseda Farms & Country Market

7281 Logerquist Rd., Baileys Harbor; 920-839-2222;
Wasedafarms.com

You may have eaten this family-owned five-hundred-acre farm's
grass-fed organic beef, pork, and chicken (plus eggs) without know-
ing it—because it is on menus at Stone Bank Farm Market near
Milwaukee, Trattoria Stefano in Sheboygan, and Trixie's and Wickman
House in Door County. A cute red barn on the farmstead—which
gifts four-year scholarships to agriculture students—sells everything
they produce, which appeals to Door County vacationers who wish
to stock the fridge or grill out. Visitors are welcome to stroll the
farm, gardens, and trails.

Wilfert Farms

7528 Manitou Dr., Two Rivers; 920-683-3264; wilfertfarms.com

Although when Joseph Wilfert founded the farm in 1877, he raised
chicken, cows, and pigs, during the 1990s the family transitioned into
growing vegetable crops. During strawberry season, you can pick
berries at Dave and Terri Wilfert's farm, but the family also grows
and sells a variety of vegetables and fruits. This includes asparagus,
watermelon, sweet corn, cabbage, broccoli, pumpkins, and tomatoes.

© Manitowoc Area Visitor & Convention Bureau

All are sold at the on-site market. During the fall, visitors enjoy the farm's pumpkin patch.

Farm Stays

Folk Tree Farm
1394 Michigan Rd., Washington Island; 920-342-5679

Bookable through Airbnb, this twenty-one-acre farmstead is owned by newer farmers Shawn Murray and Casey Dahl (who started their business in 2018 after interning at Michael Fields Agricultural Institute, followed by four small-scale Wisconsin farms) eager to share with overnight guests what they have learned. An orchard bearing fruit and nuts, vegetable gardens, a greenhouse, and crew of Oberhasli goats are on the property. Crops are woven into meals at Hotel Washington's restaurant, which helps ensure the farmers livelihood.

The G Farm's "Camping with Cows"
9328 Manu Rd., Larsen; 920-268-2856; theg.farm

Justin Duell, owner of The G Farm, rents out a tiny house dubbed "Camping With Cows" on the twenty-seven-acre property that can accommodate between two and three guests. On the property is a three-and-a-half-acre pond open for catching pan fish (walleye and perch); the pond also features docks.

The farm stay, bookable through HipCamp.com, is definitely rustic but for those who are interested in tiny-house living this is a low-impact way to test drive bunking in a compact space. "We have an outhouse so you can stay close to nature on your stay," says Duell. "There is not a shower or a bathroom, but feel free to jump in the pond."

Duell leads a farm tour lasting between thirty minutes and an hour for those who want to understand day-to-day operations of the farm more deeply. Beef, rabbits, turkeys, pork, and pasture-raised chickens are the farm's focus, with an even keener focus on farming sustainably. "You will walk away understanding how a diverse agricultural system works and why it is important," says Duell, who signs his emails "Farmer Justin." "The tour includes a visit to each of the animal groups on the farm. In the spring there are young animals and as the season follows along you are able to see production as the year progresses."

In addition to the tiny house there is a campsite. "Campers are welcome to use the canoes, paddle boat, fishing equipment, grill, and campfire on their stay," says Duell. "We try to make it as easy as packing a duffel bag and skipping town for a couple of days. We are a direct-to-consumer farm and have an on-farm store. Stop by upon arrival and pick up your evening's dinner for the grill." Indeed, the farm—which was established in 2014—store is open to anyone, not just overnight guests, seasonally and on Saturdays and Sundays. The farm also offers a meat CSA (of beef, pork, and chicken) with an option to tack on eggs. Meat cuts include bratwurst, bacon, sirloin steaks, porterhouses, ribs, rump roast, sausage, and ground beef, plus whole or half chickens.

Each winter the farm hosts an ice-fishing tournament for kids.

Sabamba Alpaca Ranch and Bed and Breakfast
2338 Hickory Rd., DePere; 920-371-0003; sabambaalpaca.com

This is not a farm where you see alpaca only through a gate or fence: Sally and Tom Schmidt want you to play with their sixty furry friends. "You can feed, walk, and interact with the herd," says Sally. It is also

© Sabamba Alpaca Ranch and Bed and Breakfast

okay to read a book on the porch with alpacas in view out at pasture. Guests at the twelve-and-a-half-acre farm stay in one of two one-bedroom suites. Each has a private bath and queen bed; one features a whirlpool. Want to learn more about alpacas? Tack on a day-long tour.

Tipping Bucket Farm
987 Townline Rd., Washington Island; 920-847-2400 or 920-535-0077; breadandwaterwi.com
With the same owners as Bread & Water—a coffee and breakfast/lunch spot on the island—this fifty-five-acre farmstead with a whimsical name invites you to try on the farming lifestyle, alongside goats, a peacock, chickens, and a pot-bellied pig. There are two farmhouse accommodations: The Artist's Studio and The Farm Apartment. Artists and writers flock here due to the inspiring landscape and the fact that this is the quieter part of Door County, with fewer tourists willing to hop a ferry.

Velvet Sheep Farms
W1681 Garton Rd., Sheboygan; 920-547-0100; Velvetsheepfarms.com
"All of our animals have names," says Kelli Scharff, who owns the seven-acre Velvet Sheep Farms with her husband, Joshua Lundgren-Dolan. She ticks off the names of some of their goats and sheep like they are her own kids: Annie Oakley, Marilyn Monroe, Mad Max, and Frank Sinatra.

Both vegans and huge animal lovers who did not grow up in Wisconsin, the couple was brought here from Dallas by Kelli's fashion-related job at Kohl's corporate headquarters but they met in Cincinnati. "It is too hot for wool-producing sheep in Texas," Kelli says, which crushed their sheep-farm dreams shortly upon arrival in the Lone Star State. That they had to live outside of the city for their jobs was another blow.

Josh, who spins yarn and learned to knit and crochet at a Waldorf school in ppstate New York, came up with the idea to own a sheep farm that ties into Kelli's knitting and fashion design. Sheboygan is where they found their bliss in 2018, transforming a former

berry farm and cornfield into a sheep and goat farm (culled for their fibers) with a bed-and-breakfast. Skeins of yarn—both hand-dyed and natural—are sold through the farm. A recently planted apple orchard hopes to bear a—pardon the pun—fruitful harvest within the next few years. They also sell raw fibers after each springs' sheep shearing. Another goal is to sell products, accessories, and apparel (such as hats, gloves, tapestries, and home décor) handmade out of those fibers. Kelli has noticed a trend at New England farms to host workshops in knitting and weaving, something she hopes to bring to their farm soon.

Being in a region that attracts luxury-minded travelers to the nearby company town of Kohler (particularly the Kohler Waters Spa and The American Club's dining and accommodations) as well as Sheboygan's "cute downtown," says Kelli, has been helpful. "Everyone is so supportive of the growth, and of us," says Kelli.

Accommodations are four bedrooms with private baths and include breakfast.

"Staying with us will mean you can look out your bedroom window in the morning to see our sheep and goats greeting you 'good morning,'" says Kelli. "They all love to be spoiled by visitors so you can spend as much time with them as you would like!"

Wellspring Education Center & Organic Farm
4382 Hickory Rd., West Bend; 262-675-6755; wellspringinc.org
One of the Milwaukee area's first vegetable farms with a CSA program—dating back to 1988—Mary Ann Ihm's thirty-six-and-a-half-acre

farm now farms on six certified-organic acres. Operating as a non-profit, seasonal staff and local experts love to share those experiences via workshops and classes (from fermented foods to beekeeping), seedling sales, and chef-led dinners (such as "Taste of Wellspring" in September). Since 1996, overnight accommodations are in two private rooms or a cottage, with shared baths and a farm-to-fork breakfast.

Farmers Markets

Algoma Sunday Farmers Market
Legion Park, 620 Lake St., Algoma; visitalgomawi.com
You will find this farmers market across from the lighthouse and Lake Michigan's shoreline in downtown Algoma. A variety of items are sold by farm vendors, such as Wery's Hog Farm, Theys Orchards (family-owned plum and apple orchard since 1941), Folkvangr Farms' microgreens, and Sap Happy Maple Products' maple syrup (owned by fourth-generation maple-syrup producers). When: Sundays from 9:30 a.m. to 1 p.m., late June through late October

Antigo Farmers Market
Peaceful Valley Park, 420 Field St., Antigo; 715-216-1846; antigomarket.com
To sell at this market, farmers must have grown what they are selling from within fifty miles. This includes eggs, vegetables, maple syrup, bread, fruit, grass-fed beef, and honey. Non-edible gifts like soap, potted plants, quilts, and all-natural beauty products—for friends and family, or maybe even yourself—are also sold at the market. Shoppers can also order prepared Asian foods for a quick lunch. When: Saturdays from 8 a.m. to noon, early June through late November

Baileys Harbor Farmers Market
Outside Baileys Harbor City Hall, 2392 County F, Baileys Harbor; 920-839-2366; baileysharbor.com
Vendors sell fruit, olive oil (from a Green Bay retailer), floral bouquets, canned fruits and vegetables, and vegetables. There is also

a henna artist; numerous artists, authors, and photographers; and alpaca-fiber apparel (from a boutique in Fish Creek with its resident alpacas). Want to give your garden a boost? Pick up plants and flowers from Salzsieder Nursery. When: Sundays from 9 a.m. to 1 p.m., late May through late October; indoor market at Baileys Harbor Town Hall's Auditorium, early November through mid-April

Downtown Appleton Farm Market

College Avenue between Appleton and Drew Streets, Appleton; 920-954-9112; appletondowntown.org

From a signature Wisconsin dessert (Uncle Mike's Bake Shop kringles) and Caprine Supreme goat cheese to yarn from Hidden Valley Farm & Woolen Mill's sheep, this market only allows products hand-crafted or hand-grown. You will also find gluten-free eats from Happy Bellies Bake Shop as well as fruits and vegetables grown at family farms in the area. When: Saturdays from 8 a.m. to 12:30 p.m., mid-June through October (vendors move to City Center Plaza, 10 E. College Ave., Appleton; Saturdays from 9 a.m. to 12:30 p.m., November and December)

Chilton Farmers Market

Klinkner Memorial Park, 815 Memorial Dr., Chilton; 920-849-2451; chiltonchamber.com

In a new location as of 2019, what is nice about this market is that you do not have to worry about oversleeping or arriving during a tight window: instead, you have five hours mid-day. Vendors sell crafts, fresh vegetables, bakery items and more. When: Fridays from 11 a.m. to 4 p.m., mid-June through late October

Downtown De Pere Farmers Market

George Street Plaza, between Broadway and Wisconsin Streets, De Pere; 920-403-0337; definitelydepere.org/downtown-de -pere-farmers-market

Scoop up artisan bread loaves from vendors like Great Harvest Bread Company and Breadsmith—both have stands at this market—or vegetables and fruits (from cantaloupe to watermelon) from Green Bay's Sunny Hill Farm. When: Thursdays from 3 to 8 p.m., late June through late August (until 7 p.m. in late September)

Downtown Fond Du Lac Farmers Market
Main Street between Forest and Western Avenues, Fond du Lac; downtownfdl.com

This market attracts about eighty vendors on Saturday with a smaller version (fifteen vendors) on Wednesday. About three thousand shoppers come to the Saturday market each week, in search of art, pottery, and crafts to supplement staples like fresh vegetables, fruits, cheeses, and meats. Plants and flowers are also sold. Wednesday's market attracts local office workers in search of alfresco lunch while picking up farm-fresh groceries. When: Saturdays from 8 a.m. to noon, mid-May through late October; Wednesdays from 11 a.m. to 2 p.m., early June through late September

Downtown Kaukauna Farmers Market
101 Crooks Ave. (public parking lot), Kaukauna; 920-766-6304; cityofkaukana.com

About thirty vendors are at this market each week, selling apples, wild-caught Alaskan sockeye salmon, cut flowers, dog treats, and—of course—fresh vegetables grown on local farms or in local gardens. There are also prepared-food vendors, for breakfast on the go. When: Saturdays from 7:30 a.m. to noon, mid-July through mid-October

Dundee Farmers & Crafter's Market
The Sportsman's Park, Highways 67 and F, Dundee; 920-528-8773

True to the name, this market shines a light on both farmers and crafters. Their wares include fishing lures, hand-knitted clothing, garden art, specialty meats, jewelry, maple syrup, farm-fresh vegetables and fruits, baked goods, and honey. "We require that all vendors are growing, raising, or making the items that they sell," says the market. When: Sundays from 9 a.m. to 1 p.m., July through October

Egg Harbor Farmers Market

Harbor View Park, 7809 Highway 42, Egg Harbor;
eggharbordoorcounty.org/events/farmers-market

Perched on a hill in Harbor View Park overlooking Green Bay, this market offers a beautiful water view. Off-beat items sold, in addition to fruits and vegetables grown at local farms and gardens (including Robertson Orchards and Sully's Produce), are essential oils, jewelry, and art. Local health-food store Greens N Grains also operates a stand. When: Fridays from 9 a.m. to 1 p.m., mid-May through late October

Elkhart Lake Farmers & Artisans Market

Village Square Park, North Lake and Rhine Streets,
elkhartlakechamber.com

This cute lake town an hour north of Milwaukee brings together seventy vendors for its farmers market, thoughtfully including art, too, such as hand-carved wooden dishes. Prepared foods include Rosa's Bakery's Mexican sweet breads. You can also pick up a beautiful bouquet of cut flowers and specialty jams. When: Saturdays from 8:30 a.m. to noon, early June through mid-October

Farmers Market on Broadway

Leicht Memorial Park, 128 Dousman, Green Bay;
downtowngreenbay.com

With 150 vendors, this is one of the state's largest farmers markets and includes a beer garden to rest your tired feet and listen to live music while sipping a pint. Keep tabs on the market's Facebook page for news of that week's vendors—and where you can find them. From Caprine Supreme cheese to Breadsmith's breads, and no shortage of fruits, vegetables, meats, and kitchen accessories like spices and honey, consider the week's shopping done. When: Wednesdays from 4 to 7 p.m., early June through late September

Jacksonport Farmers Market

Lakeside Park, Highway 57, Jacksonport; jacksonport.net

This market may be small, but it is within a town with lots of rural land on the Lake Michigan side of the Door County peninsula, some of it used for agricultural purposes. Live music is also a part of the market, which is in a park hugging Lake Michigan. When: Tuesdays from 9 a.m. to 1 p.m., mid-May through October

Lomira Farmers Market
Sterr Park, Lomira; 920-960-8732
From small family-owned farms like Critters Produce & Pumpkins (known for their tomatoes and pumpkins, plus potted mums) to Let's Get Caked baked goods and sweet treats, as well as new vendors like a hemp farmer, this market also supports local artists and crafters selling their wares. The local library also hosts "Storytime at the Market," and there is a three-acre pond in the park. Markets are in the pavilion, which helps prevent cancellation if it is raining. When: Thursdays from 5 to 7:30 p.m., early June through late August

Manitowoc Farmers Market
707 Quay St., Manitowoc; 920-686-6930; manitowoc.org
From succulents and floral bouquets to adorn your house or yard to a screw-art piece depicting Darth Vader, the variety of options at this market are vast. Among the small farms selling fruits and vegetables are Garden of Eaton, and prepared-food options include Paradise Food's Mexican-themed menu as well as other vendors' lattes and doughnuts. When: Saturdays from 8 a.m. to 1 p.m., May through October

Market on Military
1555 W. Mason St. (former Sears parking lot), Green Bay (summer); 1481 W. Mason St., Green Bay (winter); 920-544-9503; militaryave.org
Every item sold at this market in Green Bay's Military Avenue Business District must be locally grown, made, or crafted. This includes Shiitake Creek Mushroom Company's shiitake mushrooms, Apple Licious Orchard's dried-apple chips, Seven Feather Bison Ranch's meats, and From Above Bakery's cookie kits. When: Thursdays from 3 to 7 p.m., June through late October; Thursdays from 9 a.m. to 2 p.m., November through April

Oneida Farmers Market

N7332 Water Circle Place, Oneida; 920-869-4595;
exploreoneida.com

In addition to the items, you would normally find at a farmers market—like honey, maple syrup, fruits, and vegetables—this market with ties to the reservation sells sweetgrass, harvested in Oneida and often burned with sage for smudging. The Oneida Nation's apple orchard sells fruit come early September and a bison farmer is often one of the vendors. When: Thursdays from noon to 6 p.m., late June through early October

Saturday Farmers Market (Green Bay)

The Riverwalk Plaza parking lot, 200 S. Washington St., Green Bay;
downtowngreenbay.com

This market is so vibrant that it lasts for five hours on a Saturday, attracting seventy-five vendors who sell their arts and crafts, farm-raised fruits and vegetables, and grass-fed meats. You will also find cold-pressed sunflower oil (for cooking or using in salads), hanging baskets, culinary herbs ready to plant in the ground, and

certified-organic eggs. Key West Sea Soap Co.'s tropical-scented soap line and prepared pan-Asian foods are two other delights. When: Saturdays from 7 a.m. to noon, early June through late October

Seymour Farmers Market
Woodland Plaza by Highway 54, Seymour; 920-422-2380

Stay tuned to the market's Facebook page for news of that week's vendors and what they will be selling. Throughout the season its vendors include Yoder's Bakery (Amish turnovers); Lone Wolf Winery (maple syrup); Pink Fusion (spices); farmers selling fresh cut flowers, fruits, and vegetables; and even bath and body care products like laundry soap and cosmetics, plus works by crafters and artists. When: Tuesdays from 2 to 6 p.m., June through August

Shawano Farmers Market
Franklin Park, 235 S. Washington St., Shawano; thefreshproject.org

From Gnarly Marlin's Pickled Kohlrabi to farm-fresh produce grown by Hmong and Vietnamese farmers, not to mention Wendy and Kyle Jorgensen's yak meat, there is a lot of eclecticism at this market. You can also buy soap, honey, grass-fed beef, lamb meat, and smoked-fish products. A dozen or so artists and crafters participate in the market, too. Grab breakfast or lunch thanks to fried pies, Indian tacos, bubble tea, and more. When: Saturdays from 8 a.m. to noon, mid-June through early October

Stevens Point Farmers Market
Mathias Mitchell Public Square, 2nd and Main Streets, Stevens Point; stevenspointfarmersmarket.com

Meet Wisconsin's longest-running farmers market, extending back to 1847 on this very same square in downtown Stevens Point. Today a farmers-market association manages the market's vendors and publicity, offering a perk nearly unheard of in Wisconsin: a daily farmers market when in season. On the first Saturday of the month at 9:30 a.m. is "Chef on the Square," with cooking demos utilizing what is

fresh that week. When: daily from 6:30 a.m. to 5 p.m., May through October (Saturday is the busiest day)

Sturgeon Bay Farm & Craft Market

Market Square, 421 Michigan St., Sturgeon Bay; sturgeonbaywi.org
Located in Door County's largest city, and in the heart of its downtown, this market provides a stand for local businesses with name recognition—like Renard's Cheese and Door County Custom Meats—for convenient pick-up but also sells farm-fresh vegetables and fruits grown locally. Baked goods and hand-crafted items are also available. When: Saturdays from 8:30 a.m. to 12 p.m., early June through mid-October

Sustain Greenville Farmers Market

Greenville Town Hall, W6860 Parkview Dr., Greenville; Sustaingreenville.org
Because many of the marker's vendors, who have sold here for many years in a row, start their crops in greenhouses, this allows them to bring—for example—tomatoes to market the first week in June, say market organizers. A variety of craft vendors allow you to treat yourself or pick up gifts, such as woodwork, jewelry, and bath and body items. Prepared foods such as Thai Cuisine's Thai menu and Bakery on the Terrace's baked goods round out the offerings. When: Wednesdays from 3 to 7 p.m., late May through early October

Two Rivers Farmers Market

Central Park, downtown Two Rivers; two-rivers.org
For a town this small (Two Rivers is home to about eleven thousand people and in Manitowoc County) to host a market twice a week is unusual but warmly welcomed by its residents. Eggs, cut flowers, fruits and vegetables, plants for your own garden, and honey are sold by vendors. Last-minute gifts can be found at the market's stands for artists and crafters. When: Saturdays from 8 a.m. to 1 p.m., and Wednesdays from 1 to 5:30 p.m., early May through October

Folk Schools

The Clearing

12171 Garrett Bay Rd., Ellison Bay; 920-854-4088; theclearing.org
Founded in 1935 by Chicago landscape architect Jens Jensen, whose projects include Ravinia Music Festival grounds in Highland Park, Illinois, this folk school originally focused on soil-based projects. Day, two-day, and week-long classes run nine months out of the year (January and February and May through November), teaching skills in furniture making, weaving, metal work, paper arts, painting, music, quilting, birding, and wood working and carving. Instructors and students live on the one-hundred and twenty-five-acre campus and enjoy communal meals.

Food Trails

Wisconsin Coastal Food Trail

wisconsincoastalfoodtrail.com
Although Manitowoc County's artisan-food businesses, purveyors, and producers have been going strong for decades, not until 2018 were they linked together. Included are creameries (Cedar Crest Ice Cream, Pine River Dairy, and Henning's Cheese), Wilfert Farms' strawberry fields, fisheries (Susie-Q Fish Company), farm-to-table meals at places like Holla and Farm Wisconsin Cafe (an experiential farming museum and restaurant). Wineries are along the trail if they grow their own grapes as Cold Country Vines & Wines does. There are even two new Manitowoc breweries: Sabbatical Brewing on a former mill site and PetSkull Brewing Company.

Pick Your Owns

Cherry Crops in Door County

wisconsincherrygrowers.org
Nowhere else in Wisconsin—or the world—do Montmorency tart cherries grow this prolifically. That is because of the late start to

spring, being this far north. Another reason: the rocky, shallow soil type. Still another benefit is that the peninsula is wedged between Lake Michigan and Green Bay, buffered from frosts that would kill off the fruit trees' blossoms each spring.

You could spend a week in Door County and still not have sampled every orchard's cherries. Just like grape vines in Napa, anyone who can grow cherries does. Fortunately, you do not need to be here only in late July through mid-August when cherries ripen to experience that tart sensation on your palate. Farm stands sell frozen cherries year-round to ensure a cherry pie can always be baked at home. Dried cherries are sold in little packets—or in bulk— at many retail shops in the area. One of Door County Coffee & Tea Co.'s coffee flavors? Cherry Crème. And you would be hard-pressed to find a restaurant in Door County not dishing out cherry-centric desserts during the late summer and early-fall months. It is practically unheard of for a fish boil, in fact, to not close out your meal with cherry pie.

Fish boils—a Scandinavian tradition of Lake Michigan whitefish, potatoes and onions boiled in a large metal kettle over an open fire outdoors, with lots of salt—are just as much of Door County's heritage as cherry orchards. Fish boils were brought to Door County by immigrants in the late 1800s. Around that same time, the area's first cherry trees were planted.

Many local wineries—such as Door Peninsula Winery—buy cherries from local orchards to make wine. Cherry Mimosa Wine combines those cherries with apple cider for a new twist on a brunch beverage. Another example is Milwaukee's Great Lakes Distillery's Good Land Door County Cherry Liqueur.

Among the orchards that offer you-pick cherry dates in late July through early August are the following: Alexander's Cherry Orchard (Brussels), Cherry Lane Orchards (Sturgeon Bay), Choice Orchards (Sturgeon Bay), Hyline Orchard (Fish Creek), Kielar Akers Orchard (Sturgeon Bay), Meleddy Cherry Farm (Sturgeon Bay), Paradise Farms Orchard (Brussels), and Robertson Orchards (Sturgeon Bay). A more complete list is on the Wisconsin Cherry Growers' website,

Island Harvest

1249 Aznoe Rd., Washington Island; 920-847-2963

A roadside stand with a gingham-check tablecloth and on the honor system at this strawberry farm adds to the quaint factor of Washington Island, a rural community with more open land than people. Keep an eye on the farm's Facebook page for news of what hours and days the stand is open, as well as you-pick dates (limited). You may even get wind of a half-off special this way, ensuring that—once frozen—you will have access to strawberries during winter. Strawberries are grown using organic principles.

Tree Farms

County H Tree Farm

8674 County Road H, Sturgeon Bay; 920-825-7618; countyhtreefarm.com

Three generations of the same family work at this tree farm, which harvested its first trees in 1999. After noticing couples popping the question on site, fresh-cut trees are now half-off to those lovebirds. Visitors can cut their own tree and pets are welcome. Sold in the stone-walled gift shop (with its 1904 horse-drawn sleigh and a farm tree adorned with antique ornaments) are pre-cut trees, embroidered dish towels, hand-knit wool mittens and more; plus, free hot chocolate and candy canes. Children can drop letters to Santa in a dedicated mailbox.

Ottman's Fir Farm

9248 County Road A, Fish Creek; 262-719-0489; ottmansfirfarm.com

At this farm (in the same family since 1946) you can either cut your own or grab a pre-cut tree—choose from Douglas fir, Fraser fir, white pine, Canaan fir, white fir, spruce, and Scotch pine. Free hot chocolate and cider, plus cookies and candy canes, inject a festive ambiance in the days leading up to Christmas. Make sure your phone or camera battery is charged: The Christmas sleigh is a popular prop for visitors to the farm.

Cold Country Vines and Wines

E3207 Nuclear Rd., Kewaunee; 920-776-1328;
coldcountrywines.com

Traveling to Jay and Kay Stoeger's estate winery's remote setting—
just past the curtain of cold-hardy grape vines, spanning sixteen
acres with Petite Pearl, Marechal Foch, Louise Swenson, and Marquette, among others—is worth it once you settle in on the patio

with a glass of wine, paired with snacks like cheese platters (you can also bring your own food). This is Northeast Wisconsin's largest vineyard, within the Wisconsin Ledge AVA. Check the website for news of live music on weekends (Jay is a former Nashville singer-songwriter) and Harvest Fest in late September.

Door 44 Winery

5464 County Road P, Sturgeon Bay; 888-932-0044; 44wineries.com

It is impossible to miss the tasting room along Highway 42. The gleaming building is perched on the hillside, turquoise Adirondack chairs overlook the vines of this estate winery. The tasting room's bar is smack in the center—just like a Napa winery's may be—surrounded by wine retail items as well as a cooler filled with cheese curds and other artisan foods. Wisconsin natives Steve Johnson and Maria Milano—who met in law school—planted their first grapes in 2005, focused on cold-hardy varietals.

Door County Wine Trail

doorcountywinetrail.com

Bolstered by the throngs of visitors from throughout Wisconsin as well as Chicago and the Twin Cities between Memorial Day and Labor Day, winery tasting rooms in Door County have no problem attracting taste-testers. And due to the peninsula's status as a place where multiple generations in the same family, as well as couples, like to retreat, there is no shortage of wine drinkers.

But the quality of these wines is also up to snuff.

Fueled by its fortunate position on a peninsula between Green Bay and Lake Michigan, this area lends itself nicely to growing grape vines. Access to fresh fruit such as cherries and peaches has also allowed for fun blends. The first winery opened in 1974 (Door Peninsula Winery in Sturgeon Bay) and others continue to follow, with one of the newer ones being Harbor Ridge in Egg Harbor, a place under new ownership but continuing the tradition of making wine with estate-grown grapes.

The Door County Wine Trail is an attempt to create an itinerary for oenophile travelers, providing must-see and -do stops that are entirely about wine, whether it is bocce ball and tapas at Stone's

Throw Winery or chilling in an Adirondack chair with a vineyard view at Door 44 Winery. Eight wineries are on the trail: Door 44 Winery (Sturgeon Bay), Door Peninsula Winery (Sturgeon Bay), Harbor Ridge Winery (Egg Harbor), Lautenbach's Orchard Country Winery (Fish Creek), Red Oak Winery (Sturgeon Bay), Simon Creek Vineyard & Winery (Sturgeon Bay), Stone's Throw Winery (Baileys Harbor), and von Stiehl Winery (Algoma). These are not just winemakers but also vintners, keeping a close eye on the grapevines they tend. It is not uncommon for these wineries to also import grapes from other wine regions around the US to offer more "familiar" wines like Cabernet Sauvignon or Pinot Noir whose grapes thrive in other climates, not here.

Subscribe to the wine trail's newsletter via its website to stay in the know about new-release wines, recipes to cook when you open your bottles, and what is happening at the tasting rooms.

Door Peninsula Winery
5806 WI-42, Sturgeon Bay; 920-743-7431; dcwine.com
In the tasting room's a mix of old and new—via the 1868 two-room schoolhouse and a new distillery (Door County Distillery). Established in 1974 as Door County's first commercial winery, a portfolio of around sixty different wines is mostly fruit vines (such as the popular "Sunset Splash," marrying cherries and apples), although some blend with traditional grapes (like Blackberry Merlot or Strawberry Zin). The gift shop features some of the best home-décor goods in the entire county, especially if you are a foodie or oenophile who likes entertaining. Tours of the winery and distillery are held daily.

Harbor Ridge Winery
4690 Rainbow Ridge Ct., Egg Harbor; 920-868-4321; harborridgewinery.com
In 2019 Chris and Betsy Folbrecht—along with Betsy's parents: Terry and Deb Radloff—bought the ten-year-old winery, which produces twenty different wines, including some aged in hand-coopered Garbellotto barrels from Italy. Cheese wedges and soap bars bump up against wine bottles in its shop and tasting room. The recently planted vineyard of 350 Marquette grapevines—used for the

winery's "Cracklin' (sparkling) Rosé," fortified Port called "Harbor" and a blend dubbed "Crimes Against Vines"—is just outside the tasting room's door. Live music every Sunday during summer.

Lautenbach's Orchard Country
9197 WI-42, Fish Creek; 920-868-3479; orchardcountry.com
Part winery, part cherry orchard (a whopping one hundred acres)—in addition to hosting a robust farm-stand setting along Highway 42—this family business has been going strong since 1955. Under their own wine label, traditional grapes are woven in (like Gewurztraminer and St. Pepin), sourced from an estate vineyard as well as around the US along with unique blends like Lauren Elizabeth (La Crescent grapes with Honeycrisp apples). Cherry pie fillings, jams, and pitted or dried cherries are among items sold.

Visitors Bureau

LedgeStone Vineyards & Gnarly Cedar Brewery
6381 Highway 57, Greenleaf; 920-532-4384;
ledgestonevineyards.xyz

Fifteen miles south of Green Bay, the winery/brewery (the brewery launched in 2019) hosts events like downward dogs among the vines at sunset. Wine labels are veritable works of art (everything from illustrated birds to pineapples and American Indian chiefs) and made with Washington- and California-grown and estate fruit (cold-hardy grapes like Frontenac Blanc). Thursday Night Wine Down attracts locals for wine, beer, and food as they tune into live-music outdoor shows. Attracting a younger, food-driven audience, the winery also sells tees and hosts events like a grilling class and food trucks.

Parallel 44 Vineyard & Winery
N2185 Sleepy Hollow Rd., Kewaunee; 920-388-4400;
44wineries.com

This sister winery to Door 44—both are named for being on the 44th parallel—flaunts a beautiful patio spot on which to taste its wines. It also produces a rare wine for Wisconsin: ice wine crafted from St. Pepin grapes. Another unique aspect is that every wine

is born out of estate-grown grapes, cold-hardy varietals like Petite Pearl and Marquette, both of which are blended for the quirkily named "Frozen Tundra Red." Learn more on a forty-five-minute vineyard and winery tour or a two-hour Wine Sensory Experience that purports to change how you taste wine.

Red Oak Vineyard
5781 WI-42, Sturgeon Bay; 920-743-7729; redoakwinery.com
At this winery's new tasting room (as of summer 2020) sip through its portfolio right in the barrel room and taste for yourself its unique German-style wine-making methods, which have won awards at competitions. The winery produces red, white, fruit, and dessert wines, utilizing both Door County fruit and grapes grown in hotspots like Lodi, California; Yakima, Washington; and Lake County, California. The result is everything from a heavy-on-the-citrus Sauvignon Blanc to sultry Cherry Port.

Simon Creek Vineyard & Winery
5896 Bochek Rd., Sturgeon Bay; 920-746-9307;
simoncreekvineyard.com

With an impressive line-up of live music (from classic rock to singer-songwriters who have performed outside of Wisconsin), an outdoor deck and pond, and thirty-minute tours each afternoon of the operation, there are a lot of reasons to visit this winery. Founded by the late retired Army Colonel Tim Lawrie, he was a frequent commentator on national television. The winery's awards are numerous, clocking in at around two hundred medals thus far, and a testament to its quality, whether it is "Alexa Cerise" (folding in Door County cherries) or more traditional wines like Cabernet Sauvignon (non-estate grapes).

Silarian Vineyards
11715 Hilltop Rd., Reedsville; 920-242-8379; silarianvineyards.com

With a five-acre vineyard that celebrates cold-hardy grapes like Verona, LaCrescent, Oberlin Noir, and Marquette, this is the dream of Jeremy and Brenda Haese, a young couple who opened their tasting room with its lime-green walls in 2018, having purchased their farmette five years prior. More of a beer drinker? Beer from PetSkull Brewing in Manitowoc is also poured by the glass. Live-music concerts are also performed here regularly, in a picnic-like setting. Check the website for details each season.

NORTHWEST WISCONSIN

Land is a precious resource to anyone who farms in this part of the state, where rural living is nothing new, and locals are just as eager for farm-fresh food as their urban counterparts. Chefs, too, are on board with the farm-to-table trend and you will find locally raised meats, vegetables, and cheese on local restaurant menus. At the same time, cities like Eau Claire—fueled by Grammy Award-winning musician Justin Vernon, of Bon Iver, whose Oxbow Hotel puts out a delicious food spread in its restaurant—have stepped up to support these farms.

One trend percolating here for some time—and just now catching on in Southern Wisconsin—are pizza farms. No, this is not a pizza growing in the fields (we wish!). It is when a farm bakes its pizzas in a wood-burning oven outdoors, topping with as many ingredients as possible that are grown on that farm. This includes vegetables, meat toppings, tomato-based sauce, and cheese. For some, even the grains that go into the crust originate on site. You will find five pizza farms in this part of the state, including A to Z Produce and Bakery, which dates to 1998 and whose eighty-acre farm was among the first to serve pizzas.

And no corner of Wisconsin is without its creameries. Marike Gouda's farmstead in Thorp includes the creamery, cafe, and gift shop, with a keen focus on crafting the kind of Gouda you would expect in Holland (which is no joke as owner Marieke Penterman is a Netherlands native).

Cheese

Marieke Gouda

200 W. Liberty Dr., Thorp; 715-669-5230; mariekegouda.com

Marieke Penterman moved to Northern Wisconsin from her native
Holland when her husband got a job milking cows. Frustrated at
Americans' attempts at Gouda, she finally learned to make her own,
winning national and international awards within the first few years.
Dine at the farm's Cafe DUTCHess (breakfast and lunch only, such
as grilled chicken smothered in Marieke Gouda or pancakes infused
with Marieke Gouda Foenegreek). Tours can be either self-guided
(view the parlor and cheese-making process through windows,
weekdays only) or a ninety-minute experience ending with cheese
samples.

County Fairs

Ashland County Fair

Ashland County Fairgrounds, Marengo; ashlandcofair.org

This is a fair, like many Wisconsin fairs, which is entirely about
showing off locals' talents, whether that's frying up turkey legs for
sale or singing in a karaoke contest. It has been in Marengo since
1963. Animals are shown in rabbit barns, horse barns and dedicated
areas for sheep and cattle. Crafts by adults and 4-H youth are also
exhibited and kids love the carnival rides. Held: mid-August

Buffalo County Fair

Buffalo County Fairgrounds, 400 N. Harrison St., Mondovi;
buffalocountyfairwi.com

Costing nothing to enter, this annual celebration of farm life includes
deep-fried cheese curds served by the Mondovi Lions Club, char-
coal chicken courtesy of the Mondovi Conservation Club, live music
on Friday and Saturday nights, carnival rides Thursday through
Sunday, and a tri-state tractor pull Friday night at the grandstand.
Animal viewings are a true highlight, with rabbits, poultry, beef, dairy,
goat, and sheep. Carnival rides are also at the fairgrounds. Locals'

arts and crafts are displayed in the Crops and 4-H buildings. Held: last weekend of July through early August

Eau Claire County Fair

Eau Claire County Fairgrounds, 5530 Fairview Dr., Eau Claire; 715-579-4703; eauclairecountyfair.com

Focused on youth and their exhibits—whether its skills in husbandry or arts and crafts—this fair is all about community. To that end, there is no carnival. But there is an expo building, three barns showing animals the Blue Ribbon Day Kids Club. There is also a horse and pony show in the horse arena as well as a dog-obedience show. Furthering those community roots is an opportunity to add bricks—in memory or honor of a loved one—to the Tribute Garden. The fair, which started in 1924, is free to enter. Held: late July through early August

Jackson County Fair

Jackson County Fair Park, 388 Melrose St., Black River Falls; 715-284-4558; jacksoncountyfairwi.com

Food items like fry bread and FFA Alumni milkshakes are huge draws to this "old-fashioned feel" fair, say its organizers. Bands perform in the beer garden each evening while a tractor and truck pull, as well as demo derby with combines, get the crowd roaring. An antique-tractor display and carnival rides are annual highlights. Animals are exhibited in dedicated barns for horses, small animals, livestock, dairy, and swine, and arts and crafts on are display, too. Held: first weekend in August

Northern Wisconsin State Fair

225 Edward St., Chippewa Falls; 715-723-2861; nwsfa.com

The local Optimist Club's "Optimist cheese curds" are practically a tradition for fair goers as are bacon cheddar corn dogs. In addition to animal showings, a beer and wine competition, Lego-build contest, live music on the Leinenkugel stage from big-name acts like Grand Funk Railroad and Trace Adkins, and a llama costume show are on the schedule each year. The carnival boasts twenty-eight rides and the fair attracts about one hundred thousand people each

year, with exhibitors coming from three states and even Canada. Held: second weekend in July

Oneida County Fair

Pioneer Park, Kemp and Oneida Streets, Rhinelander; 715-360-9823; ocwifair.com

Dating back to 1896, this fair does not charge admission. Animals are exhibited and there are quite a few family-friendly activities from face painting to barnyard-activity shows. The fair also features a beer garden, Saturday farmers market, live music, and food that includes smoked-brisket poutine, a Friday fish fry, fried cheese curds, and even a funnel-cake-inspired latte. Local creators demonstrate skills in floral container-arranging displays and creating "the ugliest lamp." While in Pioneer Park, check out the historic railroad depot (1892) and early-1900s schoolhouse. Held: late July through early August

Pierce County Fair

Pierce County Fairgrounds, 364 N. Maple, Ellsworth; 715-273-6874; co.pierce.wi.us

Hosted in Ellsworth, the cheese-curds capital of Wisconsin (a proclamation declared by a former governor), this fair aims to cater to young and old alike with carnival rides (Ferris wheel and Tilt-A-Whirl), commercial vendors in three areas, animal exhibits and judging, and free entertainment that may include live music, a magician, or a hypnotist. Favorite foods each year are 4-H malts, hot-beef plates, and pork chop on a stick. Held: mid-August

Taylor County Cooperative Youth Fair

Northeast corner of Highways 64 and 13, Medford; 715-748-3348; witaylorcountyfair.com

A unique event is Cans for a Cause, inviting locals to create a "can sculpture" where the non-perishable cans of food are donated to local food pantries. Held: late July

Waupaca County Fair

602 E. South St., Weyauwega; 715-281-3822; waupacacountyfair.org
Musicians perform on the Frog Country Stage and kids scream on
the carnival rides (including the Ferris wheel) while more laid-back
activities are in the commercial pavilion (various vendors) and
viewing animals such as dairy, beef, swine, and sheep. Fair food
includes Farm Bureau burgers, 4-H malts, and Carnival Pizza. Held:
late August

Dairy Centers

Feltz's Dairy Store

5796 Porter Dr., Stevens Point; 715-344-1293; feltzsdairystore.com
On a forty-five-minute tour of this fifth-generation family farm, you
will see the robotic-milking facility and the dairy barn before the
grand finale: a scoop of King Cone ice cream, stemming from a Central Wisconsin creamery. Come mid September, a hay-bale maze,
small-animal petting zoo, and pre-picked (or pick your own) pumpkins debuts. The dairy store (open since 2017) is also a destination
to load up on groceries: Ruby Coffee Roasters coffee beans; cheese
curds; the farm's Black Angus beef sticks, bratwurst, breakfast
meats, and bottled milk; Wisconsin wines; and Wisconsin cheese.

Mann Valley Farm

129 S. Glover Rd., River Falls; 715-821-0020; uwrf.edu
Acting as an incubator for tomorrow's farmers who are students at
the University of Wisconsin-River Falls, this 475-acre farm—two and
a half miles north of River Falls—opens itself up to the public with
self-guided tours. Each semester between ten and twelve students
are hired on the campus farm. Per a handy brochure on the website,
you will see the milking center, lots of animals who live on property
(swine, sheep, and cows) as well as a sophisticated feed-processing
and composting setup (maybe it will inspire your own garden?).
There is also an indoor riding area used for horse shows and the
students' use.

A to Z Produce and Bakery

N2956 Anker Ln., Stockholm; 715-448-4802;
atozproduceandbakery.com

Wheat grown on Robbi Bannen and Ted Fisher's eighty-acre farm
and ground in a stone mill fuels Tuesday's pizza nights spring through
fall, baked in the same brick ovens as bread. They have hosted pizza
nights since 1998. When you order one of their pizzas, you can be
assured that nearly everything came from the farm and if not, then
nearby. This includes the farm's pasture-raised pigs, sheep, and
cattle, as well as a plethora of vegetables.

Blue Vista Farm

34045 S. County Highway J, Bayfield; 715-779-5400;
bluevistafarm.com

My favorite jam? Raspberry ginger, bought in the barn boutique
at this fruit farm boasting Lake Superior views two miles from the
Apostle Islands' shoreline. I spread that jam on toast nearly every
morning, dropped it into dough for thumbprint cookies. You-pick
"eco-grown" (say owners Eric Carlson and Ellen Kwiatkowski) raspberries, blueberries, and apples in season. They are so passionate

about protecting the farm—dating back to 1900 as a dairy farm—
they have sold development rights to the Bayfield Regional Conservancy. Tours, chef dinners, native-plant and medicinal-herb walks,
and farmhouse stays enhance your experience.

Borner Farm Project

1266 Walnut St., Prescott; 651-235-4906; bornerfarmproject.com
Unlike most of Wisconsin's pizza farms, which are on farms with
rolling hills, this one is tucked into an urban setting. And although it
is small (one acre), it is mighty. Pizza nights incorporating the farm's
dough and vegetables (grown without chemicals) kick off in April,
utilizing two wood-fired brick ovens, and extend through fall. A major
goal is to equip the community with knowledge about farming and
gardening, taught through classes at the farm. The farm also offers
CSA subscriptions for the vegetables it grows.

Dream Valley Farm

N50768 County Road D., Strum; 715-695-3617; dreamvalleyfarm.com
On this 143-acre sheep-dairy farm—established by Tom and Laurel
Kieffer in 1991—roam sheep harvested for their meat (from lamb
cuts to whole animals) as well as their wool (for shoe liners, chair
cushions, wool batting, spun yarn, and sheepskins). Products can be

shipped or picked up at the farm. Even after Tom's passing, Laurel remains committed to, as the farm's name suggests, keeping it energy-dependent and solar-passive. Visitors are welcome on shearing days in the spring, announced on the farm's Facebook page.

EB Ranch
N13346 490th St., Ridgeland; 715-949-1120; ebranchllc.com
This is a great example of a farm that does not specialize in just one crop or product. Instead, owners Erin Link and Robert Grundy have chosen to diversify. They raise turkeys, chickens, geese, and ducks, plus vegetables that include mushrooms.

But their pride and joy are San Clemente Island goats, a rare breed considered endangered and raised at pasture on their farm. In 2013, Link and Grundy turned their attention to these goats, raising between twenty and forty at a time. "At that time, there were only around four hundred and fifty of these goats left. With a recent census it is likely that number has doubled," says Link, perhaps due to their own efforts.

"Both sides of my family are deeply immersed in farming or some aspect of farming. I grew up playing in 'cricks,' 'helping' with chores, running around in the Wisconsin woods, and continuously being fascinated and entertained by the neighbors' menagerie of draft horses, ducks, chickens, and mules," says Link. Like many teens, she moved away at age eighteen, only to be lured back in her mid-twenties. "I couldn't wait to live a more homesteading lifestyle, and being close to family was a major perk. Ten years later and a move to a larger property, my partner and I are stewards to forty acres of land." Seven of those acres are open pasture, the rest wooded.

"This land is located less than a mile from my maternal grandmother's home, nestled in a beautiful valley that truly holds my heart with its breathtaking beauty," says Link.

Link culls the goat's milk for bars of soap (sold through a monthly subscription, much like a vegetable or meat CSA, at farmers markets, retail stores, and online) and occasionally teaches classes in the art of soap-making. Another way to access the farm is through its monthly shares of goat and poultry meat.

Teaching homesteading skills is important to Link, whether it is crafting beautiful aromatic bars of goat's-milk-soap or learning to process your own chicken or turkey. "I also work with other organizations like Farmers Union to host events like Pasture Walks, or Women Caring for the Land events," says Link.

Flyte Family Farm
W13450 Cottonville Ave., Colma; 715-228-2304; flytefamilyfarm.com

Weekends in autumn are when Adam and Carrie Flyte's century-old vegetable and fruit farm shows its vibrancy, with a corn maze, wagon rides, pumpkin patch, yard games, and petting zoo, but it is also a summertime site for farm-to-table dinners (in the farm field, with strung lights and sunflower bouquets) incorporating all that they grow. Dates for you-pick berries are in summer and the family operates fifteen farm stands, in cities such as Madison, Adams, Mauston, Baraboo, Portage, Colma, and Wisconsin Rapids (details on their website).

Green Light Farm
615 5½ Ave., Prairie Farm; 715-418-3880; greenlight.farm

Since 2016, first-generation farmers Maggie Sheehan and Ben Olson have grown three acres of forty different flower varieties. Drop by

© Green Light Farm

on a Saturday between 10 a.m. and 3 p.m. (Memorial Day through Labor Day) for "self-serve, grab-and-go flowers." DIY buckets of blooms can also be picked up. Workshops coach in growing flowers and floral arranging, while an exchange program through WOOFF (World Wide Opportunities on Organic Farms) grants weekly room and board to travelers in exchange for twenty-five hours of work. Dried flower wreaths are sold in November and December.

Huntsinger Farms
S3020 Mitchell Rd., Eau Claire; 800-826-7322; huntsingerfarms.com; wifarmtechnologydays.com

Many Wisconsinites do not know about this farm but they definitely appreciate horseradish, a condiment culled from the horseradish plant. At eight million pounds per year, Huntsinger Farms is the world's largest grower and processor of horseradish—under the Silver Springs Foods subsidiary—and family-owned since 1929. Current owners are Nancy Bartusch (founder Ellis Huntsinger's grandfather) and her two sons, Eric and Ryan Rygg (his great-grandsons).

Lake View Organic Farm
N1972 420th St., Maiden Rock; 715-222-8234; lakevieworganicfarm.com

Sarah and William Brenner love their farm so much they swapped "I dos" surrounded by sunflowers and free-range chickens in 2019. William grew up on the farm and represents the third generation. A self-serve farm store (open Thursday through Sunday), says Sarah, "gets people out for a visit, taste of the farm and a treat." Sold through the farm—and featuring farm-raised ingredients—are herbal teas, spice blends and seasonings, cold-pressed organic sunflower oil, fresh eggs, grass-fed beef, maple syrup and Maiden Rock Hemp CBD products (in partnership with a neighboring farmer).

Liberation Farmers
8762 3rd St., Almond; 715-366-2656; Liberationfarmers.com

John Sheffy's and Holly Petrillo's permaculture farm raises meat (pork, chicken, goat, and rabbit) and eggs but also offers a product harvested far from Wisconsin: coffee. Beans are directly traded from Oaxaca, Mexico, farmers befriended by the couple, who travel to each year's harvest. Keep an eye on the website for news of tours and workshops, such as shiitake-mushroom growing, fruit-tree planting, crafting soap from goat's milk, rotational grazing, and "BLTs on the farm." Adelante is the farm's cafe—in downtown Almond—and home to the coffee roastery plus events like Taco Tuesdays, Friday dinners, and Sunday brunch.

Monroe County Dairy Breakfast
608-769-7681

Always held on the first Saturday in June, this breakfast-on-the-farm event seeks to spotlight a family's passion for their dairy farm and remind people this industry is right in their backyard. It is hosted at a different farm each year and between two thousand and three thousand people arrive for a relaxing kickstart to their weekends.

North Wind Organic Farm
86760 Valley Rd., Bayfield; 715-779-3254; northwindorganicfarm.org

This farm grows and sells, in its farm store, strawberries, blueberries, raspberries, and apples—you can also pick your own berries. The store also sells items you may not expect, such as sweetgrass braids, brain-tanned deer hides, birch syrup, birch baskets, and *chaga* (a mushroom often brewed as tea or coffee). In the CSA program is an option to receive shares of only fruit jams. Since 1991, summer interns have helped with the harvest, gaining experience at this farm that operates on solar and wind energy.

Northstar Bison
1936 28th Ave., Rice Lake; 715-458-4300; northstarbison.com

A guided wagon trip lets you meet the grass-fed bison herd. Owners Lee and Mary Graese's hobby farm now has a meat-processing center and order-fulfillment center. Their retail store (222 Birch Ave., Cameron) sells other farmers meat, such as grass-fed beef, lamb, and rabbits; pork; corn-free chickens; turkeys; and elk—along with their lip balms, leather goods, bone broths, soaps, and seasonings. "Our goal is to be impacting four million acres (twice the footprint of Yellowstone National Park) annually by 2030 with our holistic, regenerative, land-management practices," says Sean Graese, the couple's son.

Rooted Acres Homestead
N8391 510th St., Spring Valley

Raising American Guinea hogs, chickens, rabbits, and sheep, Andrea Sorenson and David Holodnack make the meat—along with the vegetables they harvest, on a cute farm stand repurposed out of an old trailer operating on an honor system—available for farm pick-up.

The Stone Barn
S685 County Road KK, Nelson; 715-673-4478; thenelsonstonebarn.com

Farm owners Matt and Marcy Smith's refurbished barn is a popular nuptials site but so are pizza nights (weekends) on its thirty-five acres, a tradition began by the previous owner in 2006. "We grow organic herbs for our pizzas," says Matt, "and I make my own organic soaps that can be purchased in the gift shop." Ice cream, beer, wine, and non-alcoholic beverages can be ordered with pizzas. The farmstead dates to the 1890s and the Smiths are only the fourth owners.

Stoney Acres Farm

7002 Rangeline Rd., Athens; 715-432-6285; stoneyacresfarm.net

"The other Athens" hosts "pizza on the farm" weekend nights (April through November) at this third-generation, family-owned organic farm with a CSA program raising sustainably pastured pork, grass-fed beef, eggs, vegetables, and maple syrup. A *Wisconsin Foodie* episode highlighted how amazing the pizzas taste and a new line of frozen pizzas via home delivery ensures even greater reach. In recent years, the owners also expanded into beer, introducing Stoney Acres Brewery (ciders, stouts, bocks, and lagers).

Tavis' Wild and Exotic Mushrooms

facebook.com/mushroomtavis/

Tavis Lynch is nuts about mushrooms and loves to engage with others who are, too. His Northern Wisconsin farm grows and sells oyster, wine-cap, and shiitake mushrooms. The engagement part comes in during his guided mushroom-identification adventures, mushroom talks, and educational walks in the woods (all about mushrooms, of course). He even sells mushroom grow kits through his Etsy shop and the farm, as well as through online orders. Stay in touch about events through his Facebook page.

Wetherby Cranberry Company

3365 Auger Rd., Warrens; 608-378-4813; freshcranberries.com

Cranberries are one of Wisconsin's largest exports (plus the state's largest fruit crop) and the Van Wychen family, now on its fifth generation, operates the state's only cranberry farm where—on the first Saturday October, during harvest—you can don hip waders and walk straight into the bog. (Talk about a photo op!) Score fresh cranberries, honey, dried cranberries, and wine in the two-hundred-acre farm's store, open from early October through Thanksgiving. Since 1973, Warrens has hosted the annual Warrens Cranberry Festival. The family also sells at Dane County Farmers Market.

Farmers Markets

Bayfield Farmers Market
First Street between Rittenhouse and Manypenny, Bayfield.org
Many of the micro farms and food artisans that sell at this market
mirror Bayfield's charm, such as Nunga Nunga Farms' apple cider
and fruit-pepper jellies, Angel Acres Farm's rib chops, or Starlight
Kitchens' innovative breads (such as the signature Bayfield Apple
Cinnamon loaf). Pick up a bouquet of Boda Bayfield's fresh-cut
flowers and hand-made bars of Lake Superior Lather Soap while you
are at it. When: Saturdays from 8:30 a.m. to noon, late June through
mid-October

Bruce Community Farmers Market
**Bruce Telephone Company garden, corner of Highway 8 and North
Alvey Street, Bruce; 715-868-2885**
From produce and farm-fresh eggs grown at Le Jardin Lavaliere
(owned by market manager Mary Lavaliere) to Foragers Harvest's
birch syrup and apples, there is a little bit of everything at this farmers market, tucked into a garden setting. Blue Hills Cafe's breads and

cookies are also sold here and you can pick up meat for that night's dinner thanks to Maple Hill Farm's lamb, beef, pork, and chicken. When: Fridays from 1 to 5 p.m., early June through mid-October

Downtown Eau Claire Farmers Market

Phoenix Park, 300 Riverfront Terr., Eau Claire; 715-563-2644; ecdowntownfarmersmarket.com

Vendors at this producer-only farmers market mirror Eau Claire's new creative class, from kombucha crafted in small batches to

© Visit Eau Claire

Gracie Girl Bakery's healthy goods featured on the *Today Show*, and sell their products under an open-air pavilion. Yet you will also find culinary products from Hmong family farmers as well as Bohemian Ocean shrimp, Hooterville Junction bison, and trout from Bullfrog Fish Farm. Between November and April, the market moves to the second Saturday of the month, with shortened hours (9 a.m. to 12:30 p.m.), at the L.E. Phillips Senior Center, 1616 Bellinger St., Eau Claire. When: Saturdays from 7:30 a.m. to 1 p.m., mid-May to late October

Eagle River Farmers Market
601 Michigan St., Eagle River; 715-477-0645; eaglerivermainstreet.org
This popular lake-resort town debuted its farmers market in 2001. While you can certainly pick up fresh vegetables and fruits—along with honey, jams, syrups, bread, cheeses, and flowers—crafters sell birdhouses and other works of art to fill your home and yard. Kids activities are also part of the market, appealing to the tiniest of shoppers. When: Sundays from 10 a.m. to 2 p.m., early May to early October

Hayward Farmers Market
15886W US 63, Hayward; 715-550-5720
The organizers of this market refer to it as "a visual and aromatic feast" due, in part, to the vendors' stash. This includes seasonal fruits and vegetables that are locally grown, plus eggs, honey, maple syrup, pasture-raised meats and poultry, jams, and preserves, and cut flowers. Local food artisans tote fresh breads and other baked items for sale. You can also find nursery plants to grow in your own garden. When: Mondays from 11:30 a.m. to 4 p.m., June through September

Madeline Island Farmers Market
Across from the ferry office on Main Street at Le Seuer, La Pointe; 715-765-4298; madelineisland.com
Angel Acres Farm—a certified-organic family-owned farm in Mason producing Berkshire pork, grass-fed beef, and Cornish game hens—is

one of the vendors at this market, one of only a handful that are on an island. Madeline Island is an inhabited island in Lake Superior, off the coast of Bayfield and part of the Apostle Islands National Lakeshore. You can also find fruit (strawberries, raspberries, and blueberries) from North Wind Organic Farm in Bayfield. When: Fridays from 9:15 a.m. to noon, early June through third weekend in October

Medford Farmers Market
Saturday market: Taylor County Fairgrounds, Highways 13 and 64, Medford; Tuesday market: Whelen Street across from the post office, downtown Medford

What is helpful about this market is it is not locked into one day per week. Instead, you can shop twice per week (Saturday and Tuesday). Everything you would expect to be sold here is: including farm-fresh vegetables and fruits, cut flowers, bakery items and maple syrup, plus crafts by local artisans. Check the market's Facebook page for up-to-date info on which vendors are appearing each week and what crops are ripe. When: Saturdays from 8 a.m. to noon and Tuesdays from 1 to 5 p.m., mid-May through late October

Rome Farmers Market

Alpine Village, Rome; 608-339-6997; romefarmersmarket.com

Just because Rome in Adams County is tiny (pop. 2,720) does not mean the farmers market is small. Vegetables, fruits, honey, maple syrup, cheese, jams, jellies, and preserves are sold, along with bakery items and arts and crafts. You can even pick up fresh seafood from Supreme Seafood's truck and egg rolls from stands. The market moves indoors the first Friday of the month between December and April to Rome Town Hall, 1156 Alpine Drive, Nekoosa. When: Fridays from 9 a.m. to 2 p.m., late May through early October

Spooner Farmers Market

805 S. River St., Spooner

Joining local growers of fruits and vegetables are a local wood-worker, breadmaker, soapmaker, and a pickle preserver, as well as CBD products, essential oils, heritage pork, kettle corn, eggs, and jewelry. The market's Facebook page updates throughout the season with vendor news and what they are selling. When: Saturdays from 8 a.m. to 1 p.m., mid-June through late October

Superior Downtown Farmers Market

1215 Banks Ave., Superior; 218-349-0943

Food trucks provide a quick, tasty lunch and musicians live enter-tainment while you shop for art, fresh fruits, and vegetables, and crafts at this market in downtown Superior. When: Wednesdays from 11 a.m. to 2 p.m., late May through late October

Farm Stays

Dancing Yarrow

S193 County Road BB, Mondovi; 715-309-5238; farmtoforkretreat.com

Tapping into the state's pizza-farm trend, this boho-chic retreat (whose wedding events skew traditional and are dubbed "Un Wed-dings") serves up pizzas Thursday, Friday,and Saturday nights, the vegetable toppings culled from the gardens and meat from local farms. On Thursdays is an open mic. You can also rent kayaks. Two types of stays are in the lodge—a two-bedroom chunk or the remaining five bedrooms at once. The farm also features a one-room cabin for overnight stays or you can camp in your own tent.

EverGood Farmstay

3673 County Road A, Rhinelander; 715-610-4759; evergoodfarm.com

Jenny and Brendan Tuckey's newly renovated one-bedroom apartment—above the garage and across the street from their 110-year-old farmhouse—is bookable through Airbnb. "We welcome those interested in volunteering with prior arrangement. Guests can interact with animals if interested, and take a stroll through the fields," says Jenny. "We are also on the Sugar Camp snowmobile trail and welcome small groups of snowmobilers in the winter." Non-overnight guests can book a tour and shop for organically grown vegetables at the farm stand. The farm has been in operation since 2010.

Farmhouse: Hayloft

858 Fort Rd., La Pointe; 715-747-3276; farmhousemadelineisland.com

Getting to Madeline Island is a trek—a six-hour drive from Milwaukee to Bayfield, followed by a twenty-five-minute ferry ride—but so worth it. This is part of the Apostle Islands National Lakeshore, Wisconsin's only inclusion with the National Park Service. Accommodations are in a six-room farmhouse-style inn owned by a young couple who also run the restaurant (Gip Matthews is the chef, Lauren Schuppe's operations manager), where every meal (from "Green Eggs & Ham" to a Lake Superior whitefish sandwich) is prepared with as much local produce as possible and other foods grown by the island's farmers.

Glacial Lake Cranberries

2480 County Road D, Wisconsin Rapids; 715-887-4161; glaciallakecraberries.com

The best time of year, if you can snag a reservation, to stay in The Stone Cottage is during the harvest in early October. That is because cottage guests get a free tour of the family's cranberry farm, only offered to one other group (guests at Le Chateau—The Manor Bed & Breakfast). And this family is deeply invested: While the six-thousand-acre property has been harvesting cranberries here since 1873, they have managed it since 1931. In the one-bedroom cottage is a wood-burning fireplace, full bath, kitchen, and screened porch.

Palmquist Farm
N5136 River Rd., Brantwood; 715-564-2558; palmquistfarm.com

Jim and Helen Palmquist offer a resort-type experience on their eight hundred acres. Guests can bring their horses, ice skate on the pond, relax in a Finnish sauna, or hop in horse-drawn sleighs, and stay in one of four cabins (with wood-burning fireplaces), a room in the inn, or one of five farmhouse rooms. All have private baths. Their "like coming home to Grandma's house," mantra has been in place since the farm stay's 1949 founding, in the same family since 1902. During winter you can hunt white-tail deer or ruffed grouse.

Spur of the Moment Ranch
14221 Helen Ln., Mountain; 800-644-8783; spurofthemomentranch.org

Ann Maletzke's twelve-acre ranch offers everything from a primitive camp site to a cozy log cabin. "Get up close and personal with our equines, currying and brushing, learning to pick feed, learning about tack, and helping to put it on, leading a horse safely, going out in our large paddock area, and learning to mount, ride, trot if people are comfortable," says Maletzke, "and guide around barrels, cones, play with our gigantic soccer ball, maybe end with a trail ride."

Willow Creek Ranch
E5702 Spring Coulee Rd., Coon Valley; 608-451-2861; willowcreekfoods.com

A stay in either The Trout Lodge (sleeps up to five, surrounded by pear and apple trees) or The Summer Kitchen Cabin (sleeps two) on this organically managed, diverse sustainable family farm in the Driftless Region can be restorative. Pick up "souvenirs" from your stay—pastured pork, grass-fed beef, and chicken meat and eggs—at the farm store.)

Folk Schools

Buttermilk Falls CSA and Folk School
P.O. Box 668, 599 28Oth St., Osceola; 715-294-4048;
buttermilkcsa.com
Located at Philadelphia Community Farm—a biodynamic farm operating as a non-profit since 1989—this folk school offers artists-in-
residence positions to help foster creativity without the interruption
of day-to-day life back home. Workshops and classes span topics
like dying wool with the help of local plants, while annual events
with food, music, and other arts are centered around moments such
as the fall equinox, or simply an open-mic with ice cream for all, or
perhaps a potluck with a piano performance.

Lost Creek School for Self-Reliance and Folk Art
21740 Siskiwit Lake Rd., Cornucopia; 715-953-2223; 22475
Highway 13, Cornucopia (store); lostcreekadventures.org
Any time spent along the Apostle Islands National Lakeshore is spiri-
tual and rejuvenating but especially if you are deep in thought (and
artful instruction) at this folk school. Earth Skills courses include
basketry, edible plants, hunting and gathering, and woodcarving.
Hostel-style lodging has access to a kitchenette. That the company
is also an outfitter for kayaking and stand-up paddle-boarding day
trips (you can even kayak to campgrounds on the islands), plus day
tours like animal tracking and an edible-plants hike, means a one-
stop shop for tapping into both mental and physical jaunts.

Pick Your Owns

Hauser's Superior View Farm
86565 County HighwayJ, Bayfield; 715-779-5404;
superiorviewfarm.com
At this fifth-generation apple and cherry farm (founded in 1908
by a Swiss-American horticulturist) you are gifted with a view of
Apostle Islands National Lakeshore from its hillside perch (pro tip:
head to the red barn's hayloft for the best view). That red barn is

the product of a Sears mail-order in 1928. Pick up fruits, vegetables, hard cider from Apfelhaus Cidery, perennial plants, flowers, as well as fruit jams, butters, and jellies crafted by the family. The current generation also added renewable solar energy.

Hidden Acres Berry Farm
W3865 County Road HH, Eau Claire; 608-269-6972
You can pick your own blueberries at this farm or, if in a hurry, buy containers of pre-picked berries. Keep an eye on the farm's Facebook page for news of when that season's blueberries have ripened and when to come pick in the fields. The farm's 3,500 blueberry bushes grow nine varieties and some of those bushes are nearly three decades old.

Rush River Produce
W4098 200th Ave., Maiden Rock; 715-594-3648; rushriverproduce.com
Terry and John Cuddy's nine-acre blueberry farm opens for you-pick options on select dates, though always Thursday through Sunday during July and August, and longer into the season than many of the state's berry farms. It is about three miles off Great River Road and a short walk to a scenic overlook. Call before you come, just to be sure enough of its fourteen varieties are ripe. Also grown at the farm are gooseberries and currants (red, black, and white) and locally made honey and maple syrup are also sold.

Sunset Valley Orchard & the Apple Branch
86520 Valley Rd., Bayfield; 715-779-5510; sunsetvalleyorchard.com
For over fifty years this farm has continued the dream that Craig and Sharon Johnson had when they uprooted from their home on the south side of Chicago for a new, wildly different chapter and to raise four kids on the farm. Pop into The Apple Branch for not only ten varieties of pre-picked apples (you can also pick your own apples and grapes) but also scarves, handbags, and body scrubs for the full boutique experience.

Mommsen's Harvest Hills

1696 20½ St., Rice Lake; 715-234-2665;
ricelakepumpkinpatch.com

Growing pumpkins and apples, this forty-acre farm comes alive each
fall (September and October) with a petting zoo, caramel apples,
hayrides, a pumpkin "artillery" (pumpkins are shot through a cannon), and corn mazes. You can also pick your own apples and pluck a
pumpkin right out of the patch. It is also a site that couples book for
their wedding day and guided group tours for kids include a glass of
cider, small pumpkin, and a coloring book.

Tree Farms

Country Mission Farm

4532 N. County Road X, Mosinee; 715-572-5788;
countrymissionfarm.com

This family-owned, 109-acre, and dog-friendly tree farm—owned by
Regan and Ann Pourchot—is about more than picking out your tree
(Balsam Fir, Norway Spruce, Scotch Pine, Colorado Blue Spruce,
White Spruce, White , and Fraser Fir). It is like you are in the family's
living room. Enjoy hot chocolate next to the wood stove, pick up
holiday gifts (crafts and locally made products such as maple syrup),
and snap photos in gingerbread and snowmen cut-outs. The farm is
not solely about profit: Each year the family donates a percentage of
proceeds to a charitable cause.

Highland Trees

T8501 N. 33rd St., Wausau; 715-675-7585;
highlandtreeswausau.com

Even while selling nine types of trees—including Fraser Fir, Korean
Fir, and Balsam Fir—as pre-picked or cut-your-own, this family-run
farm also retails wreaths and garland. Come and stake out your tree
as early as September 15, then return later in the year to take it
home.

Newby's Evergreen Farms

23305O County Road J, Wausau; 715-574-7272; newbysfarm.com

This cut-your-own Christmas tree farm throws a holiday fete during the season with wagon rides (including out into the fields to pick your tree), free coffee and hot cider, and a bonfire, plus last-minute gift shopping in the boutique, which sells crafts made by local artisans.

Wine and Beer Tastings

Amery Ale Works

588 115th St., Amery; 715-268-5226; ameryaleworks.com

This familiar label to Wisconsin-beer drinkers invites you to visit the brewery's farm, whether for a yoga class, live music, trivia or cribbage nights, or a suds-and-grub experience (like pizzas or beer-brisket nachos) in the taproom, which is actually in the lower level of a barn. The brewery is open between Thursday and Sunday. Not ready to go home? The brewery rents out the barn's upper level on Airbnb.

Bayfield Winery and Seven Ponds Winery

87380 Betzold Rd., Bayfield; 715-779-3274; wineriesonbetzoldroad.com

Bayfield Winery's and Seven Ponds Winery's seven-acre vineyard (next to the tasting room and on the 142-acre farm) produces Marquette, St. Croix, Frontenac, La Crescent, Brianna, Adalmiina, Itasca, and St. Croix grapes. Occasionally Renate and Scott Hauser buy grapes, but always from within Wisconsin. Second-generation owners Ian Hauser and Caitlin Mathewson operate Seven Ponds Winery, also a mix of local and estate wines. On a tasting, order plates of cheese and meat or listen to live music (weekends). Hard One ciders (hopped and ginger) are crafted from the family's apples and Michigan hops.

Belle Vinez Vineyard and Winery

W10829 875th Ave.; River Falls; 715-426-9463; bellevinez.com

With Mediterranean-tile roofs and pizzas baked in a brick oven, you may swear you have just stepped into Italy but, nope, this is the Zimmermans' winery in western Wisconsin. When translated the winery's name means "beautiful vines," what you see on a visit. Illustrated labels are a of storybook quality, depicting a woman shopping, male golfers, a bride on her wedding day and a child walking to school. All wines are crafted from Wisconsin grapes. With two tasting rooms and the aforementioned brick oven, it is easy to linger here.

Chateau St. Croix Winery & Vineyard
1998 WI-87, St. Croix Falls; 715-483-2556; chateaustcroix.com
Designed to emulate a French château, the comfy but castle-like vibe stretches beyond towering twin stone lions out front. The dog-friendly, fifty-five-acre property includes formal gardens, stables, a vineyard, and a carriage house. Sourcing grapes from Wisconsin and California, the owners (Irv Geary, who also grows grapes at his Minnesota residence; and Adam Lamoureux) are committed to using what works best in a blend or single-varietal wine. Six varietals of grapes have grown here since 2004. Order cheese, salami, or chocolate to enjoy with your wine tasting.

Cracked Barrel Winery & Vineyard
570 Coulee Trail, Hudson; 715-690-2217; cbwinery.com
Its owners (a husband-wife team who are veterans) love to tout that the St. Croix River (a popular fly-fishing spot) is only a two-minute drive from the winery. In the vineyard—adjacent to the tasting room's patio—cold-hardy grapes thrive and are used to make the couple's wines. During cooler weather, a fireplace indoors keeps things cozy. A unique experience, especially for home winemakers who want to learn from the pros, is a customized wine-making experience that takes you from fermentation to bottling.

Danzinger Vineyards
S2026 Grapeview Ln., Alma; 608-685-6000; danzingervineyard.com
Meet David and Melvin Danzinger's retirement project: an eighteen-acre vineyard with a tasting room. The two brothers are not new to farming, however: David earned a master's degree in agricultural education from the University of Wisconsin-River Falls

and ran a dairy farm while Melvin earned a bachelor's degree from UW-River Falls in agronomy. Wisconsin grapes go into every wine, including Frontenac for a semi-sweet red called "Sunset Ridge" that the owners suggest sipping with pizza; and "November Dawn," crafted from King of the North. Live music is hosted in the tasting room on occasion.

Maiden Rock Winery & Cidery
W12266 King Ln., Stockholm; 715-448-3502;
maidenrockappleswinerycidery.com

Carol Wiersma and Herdie Baisden opened their winery and cidery in 2008. Located along the Great River Road Wine Trail, this winery welcomes dogs on the patio and the apple orchard sells trees from its nursery). You can taste wine and cider, in self-designed flights. Spanning eighty acres, during apple season pick up fourteen varieties of bagged apples. Twice per summer Maiden Rock Winery & Cidery cruises sail up the St. Croix River on a paddlewheel riverboat; other events include a holiday-harvest festival (November) and wassailing by the bonfire (December). During fall school-age groups arrive on tours.

Villa Bellezza Winery and Vineyards
1420 Third St., Pepin; 715-442-2424; villabellezza.com

Snug on Great River Road (a scenic drive also known as State Highway 35), cold-hardy varietals in the Tuscany-inspired winery's twenty-acre vineyard result in thirty award-winning wines. Derick and Julianne Dahlen planted the first vines in 2002. Do not just taste the wines: monthly Italian-cooking classes with Chef Antonio (of Il Forno, the winery's eatery) inspired by his mother's kitchen back in Italy, and forty-five-minute winery tours (followed by a tasting) are other reasons for a visit. A European-style Christmas market (between Thanksgiving and Christmas) is another draw with vendors selling wares in a piazza with twinkling lights.

APPENDIX:
Organizations That Support Wisconsin Farming

It is not uncommon for new owners of farms to move to Wisconsin from another state, drawn in by the incredible resources—both financially and emotionally—to live out their farm dreams. Here are a few examples of people who are here to help.

Association of Women in Agriculture at the University of Wisconsin-Madison

uwmadisonawa.org

A student-run organization on campus with about one hundred female members—and five hundred alumni who serve as mentors in agricultural careers—that started in 1973. The association also keeps a house on campus for twenty-four members and hosts professional agricultural events (also for members).

Dairy Farmers of Wisconsin

608-836-8820; wisconsindairy.org

Based in Madison, this farmer-owned non-profit (since its 1983 debut) markets dairy products—including cheese, milk, yogurt, and butter—crafted at family-owned creameries and helps to ensure their financial future through global recognition campaigns so the farmers can do what they do best: farm.

Family Farm Defenders

608-260-0900; familyfarmers.org

With an eye on reforming our country's food system so the one-hundred-mile diet is no longer a trend, but a way of life, and all food is grown sustainably with regards to animal welfare and farm-worker rights, this Madison-based, national grassroots non-profit was founded in 1999.

Michael Fields Agricultural Institute

262-642-3303; michaelfields.org

Based in East Troy, this teaching-based non-profit hosts interns from around the world to kickstart their farming careers and was founded in

1984 by Ruth Zinniker and Christopher and Martina Mann. The organization also lobbies on behalf of farmers, in Wisconsin and beyond.

MOSES (Midwest Organic & Sustainable Education Service)
888-90-MOSES; mosesorganic.org
An annual MOSES organic-farming conference each February presents workshops led by farmers and serves to share information among Heartland farmers striving to grow food without the use of chemicals.

Soil Sisters
soilsisters.wixsite.com
Founded by Inn Serendipity co-owner Lisa Kivirist, this networking group for female farmers in Southwestern Wisconsin hosts a weekend of tours, meals, and activities at female-owned area farms (first full weekend in August) and is a program within Renewing the Countryside (a non-profit).

Wisconsin Agricultural Tourism Association
608-235-5925; wiagtourism.com
Banding together, the members' goal is encouraging travelers to explore, eat, enjoy, and sleep at farms and agricultural destinations in Wisconsin. In September, the association hosts Porkfest Farmers Market: a day of crafts, brats, and roast-pork dinners to raise funds for area food pantries.

Wisconsin Apple Growers Association
920-478-4277; wisconsinapplegrowers.org
Representing apple orchards and those involved in processing apples as fruit or cider, the website publishes a directory of orchards open to the public.

Wisconsin Berry Growers Association
608-235-5925; wiberries.org
Dedicated to promoting and supporting the growing of strawberries, raspberries, and blueberries throughout Wisconsin, its members are both commercial farmers and hobbyist gardeners, and the purpose is to be "a voice for the berry industry in Wisconsin."

Wisconsin Cheese Makers Association

608-286-1001; wischeesemakersassn.org

Serving dairy processors and industry-supply partners around the US, the organization dates to 1891 and hosts the annual US and World Championship Cheese Contests, always in Wisconsin.

Wisconsin Cherry Growers

wisconsincherrygrowers.org

Focused on Northeastern Wisconsin, especially Door County, where the state's red-tart cherry harvest occurs, the association's members are growers, manufacturers, and raw-product processors.

Wisconsin Christmas Tree Producers Association

608-742-8663; christmastrees-wi.org

Members are dedicated to providing, each year in the weeks leading up to the winter holidays, farm-grown, fresh-cut Christmas trees, wreaths, garland, and other greenery.

Wisconsin Fresh Market Vegetable Growers Association

wisconsinfreshproduce.org

Members are both commercial farmers and gardeners, growing for profit and retail; and vegetable-gardening hobbyists, sharing resources and information.

Wisconsin Farm Bureau Federation

wfbf.com

As the statewide arm of sixty-one farm bureaus in counties across Wisconsin, this membership-based foundation lobbies at the state and national level on behalf of farmers.

Wisconsin Farmers Market Association

wifarmersmarkets.org

Managers of farmers markets around the state share information about marketing and community building, all designed around the market's operation. The association hosts a three-day conference each January.

Wisconsin Farmers Union

wisconsinfarmersunion.com

An advocate for family farms, this organization seeks out educational opportunities, civic engagement, and cooperative endeavors, all about rural agriculture.

Wisconsin Grape Growers Association

920-478-4499; wigrapes.org

Uniting wineries and vineyard owners, as well as grape growers, around the state, the association aims to protect the art, science, and commerce of Wisconsin viticulture.

Wisconsin Honey Producers Association

wihoney.org

Members are not just commercial beekeepers and honey retailers—hobbyists are also welcome. The organization dates to 1864.

Wisconsin Potato and Vegetable Growers Association

wisconsinpotatoes.com

Based in Antigo, this association's mantra is to educate its 140 grower-members about growing, harvesting, shipping, researching, and advocating for the state's potato and vegetable crops.

Wisconsin Sheep Dairy Cooperative

715-360-8552; sheepmilk.biz

Members of this Northern Wisconsin-based cooperative farm sheep sustainably on family-owned farms. In 2011, its members began to pool their sheeps' milk for a cheese called Dante.

Wisconsin State Cranberry Growers Association

715-423-2070; wiscran.org

Since 1887, the association has advocated for the state's cranberry growers, which represent 60 percent of the crop nationwide.

INDEX

ABOUT THE AUTHOR

Based in Milwaukee's Bay View neighborhood, Kristine Hansen is a nationally recognized food, drinks and travel
author with articles about Wisconsin's
cheese published on Travel + Leisure's
website as well as on Fodors.com,
CheeseProfessor.com and Shondaland
.com, Shonda Rhimes' lifestyle site.
Her articles have also appeared on
ArchitecturalDigest.com and in *TIME*
magazine, *Midwest Living* magazine
and *Milwaukee Magazine*. She is the
author of *Wisconsin Cheese Cook-*

*book: Creamy, Cheesy, Sweet, and Savory Recipes from the State's
Best Creameries.*